CONTENTS

JN034785

Chapter 1
Movements in the Info-communications Industry as a Whole

Chapter 2
Situation of Info-communications Service Usage

Chapter 3
Situation of TCA Members

Chapter 1
Movements in the Info-communications Industry as a Whole

1-1   Trends in Sales in the Communications and Broadcasting Industry

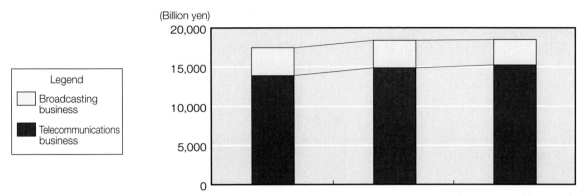

（Companies: Billion yen）

	FY2018		FY2019		FY2020	
	Number of companies	Sales amount	Number of companies	Sales amount	Number of companies	Sales amount
Communications and broadcasting industry in total	976	17,457.8	948	18,376.0	1,009	18,472.7
Telecommunications business	403	13,903.2	407	14,872.6	443	15,240.5
Broadcasting business	573	3,554.6	541	3,503.3	566	3,232.2
Commercial broadcasting	376	2,387.5	351	2,252.3	373	2,011.5
Cable television broadcasting	196	429.8	189	513.7	192	506.9
NHK	1	737.3	1	737.3	1	713.8

Note: The data on NHK are based on Published Material
       The aging comparison requires attention on the grounds that the number of valid responses varies from year to year
*Compiled by TCA based on data publicized by the Ministry of Internal Affairs and Communications

1-2 Trends in Amounts of Investment in Plant and Equipment Acquired by the Communications and Broadcasting Industry

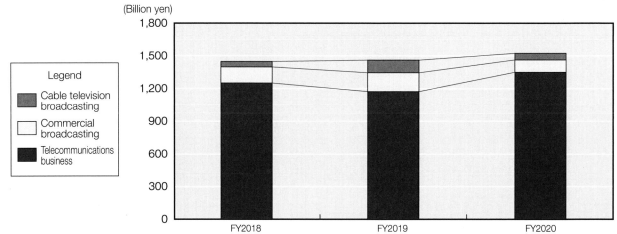

		Communications and Broadcasting Industry		Telecommunications business		Broadcasting business		Commercial broadcasting		Cable television broadcasting	
		companies	Billion yen	companies	Billion yen	companies	Billion yen	companies	Billion yen	companies	Billion yen
Amount of FY2018	Amounts of investment in plant and equipment	599	1,448.1	232	1,250.8	367	197.4	209	147.2	158	50.1
	Amounts of investment in plant and equipment (except software)	574	1,282.9	218	1,099.4	356	183.5	201	134.8	155	48.6
	Software	281	165.2	109	151.3	172	13.9	125	12.4	47	1.5
Amount of FY2019	Amounts of investment in plant and equipment	561	1,461.0	216	1,172.9	345	288.2	196	172.6	149	115.5
	Amounts of investment in plant and equipment (except software)	542	1,246.0	204	971.2	338	274.8	190	161.0	148	113.8
	Software	274	215.1	106	201.7	168	13.3	119	11.6	49	1.7
Amount of FY2020	Amounts of investment in plant and equipment	586	1,523.0	237	1,350.1	349	172.9	202	112.9	147	60.0
	Amounts of investment in plant and equipment (except software)	555	1,343.2	213	1,181.0	342	162.2	197	103.9	145	58.3
	Software	290	179.8	121	169.1	169	10.7	124	9.0	45	1.7

Note: The aging comparison requires attention on the grounds that the number of valid responses varies from year to year
*Compiled by TCA based on data publicized by the Ministry of Internal Affairs and Communications

1-3 Trends in Number of Telecommunications Carriers

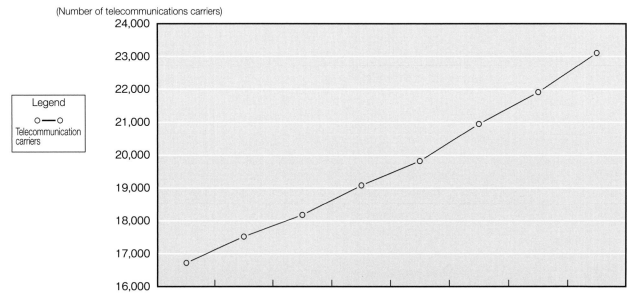

(Companies)

(Figures are current as of the end of each fiscal year.)	FY2014	FY2015	FY2016	FY2017	FY2018	FY2019	FY2020	FY2021
Number of telecommunications carriers	16,723	17,519	18,177	19,079	19,818	20,947	21,913	23,111

*Compiled by TCA based on data publicized by the Ministry of Internal Affairs and Communications

1-4   Trends in Number of Employees in the Communications and Broadcasting Industry (by Type of Service and Type of Employment)

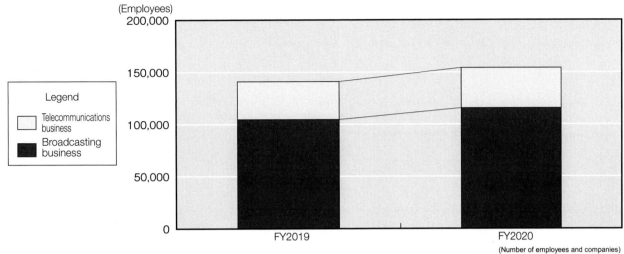

(Number of employees and companies)

	Communications and Broadcasting Industry		Telecommunications business		Broadcasting business		Commercial broadcasting		Cable television broadcasting	
	FY2019	FY2020	FY2019	FY2020	FY2019	FY2020	FY2019	FY2020	FY2019	FY2020
Number of companies	896	962	374	408	522	554	342	371	180	183
Number of employees	141,033	154,030	104,578	115,456	36,455	38,574	25,875	27,439	10,580	11,135
Number of regular employees	140,359	152,441	104,516	114,666	35,843	37,775	25,265	26,643	10,578	11,132
Permanent employees	100,840	117,799	71,523	86,358	29,317	31,441	20,662	21,968	8,655	9,473
Non-permanent employees (part-time workers, etc.)	20,193	19,236	16,227	15,168	3,966	4,068	2,634	2,975	1,332	1,093
Employees detached to other companies/organizations	13,098	12,335	12,116	11,416	982	919	718	816	264	103
Temporary employees	674	1,589	62	790	612	799	610	796	2	3
Dispatched employees	22,524	23,535	15,899	17,074	6,625	6,461	5,649	5,306	976	1,155
Number of employees per company	157	160	280	283	70	70	76	74	59	61

Note: The aging comparison requires attention on the grounds that the number of valid responses varies from year to year
*Compiled by TCA based on data publicized by the Ministry of Internal Affairs and Communications

Chapter 2
Situation of Info-communications Service Usage

2-1 Situation of Number of Contracts for Various Services

2-1-1 Trends in Number of Telecommunications Services Subscriptions, etc.

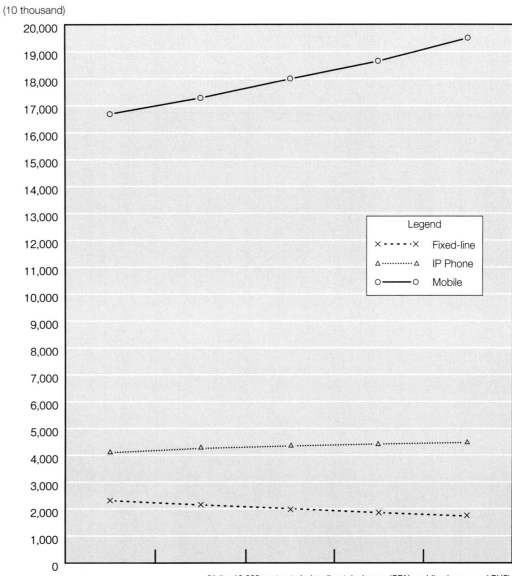

(Units: 10,000 contracts (subscriber telephones, ISDN, mobile phones, and PHS);
10,000 units (public phones); and 10,000 telephone numbers (IP phones))

Service		FY2016	FY2017	FY2018	FY2019	FY2020
Fixed-line Service Total		2,315	2,151	2,011	1,861	1,731
	Subscriber Telephone	1,987	1,845	1,724	1,595	1,486
	ISDN	312	290	272	251	231
	Public Phone	16	16	16	15	15
IP Phone		4,099	4,255	4,341	4,413	4,467
	(0ABJ-IP phone)	3,245	3,364	3,446	3,521	3,568
	(050-IP phone)	853	891	895	892	899
Mobile Service Total		16,685	17,279	17,987	18,651	19,505
	Mobile Phone	16,350	17,019	17,782	18,490	19,440
	PHS	336	260	206	162	66

Note: Figures for "Public Phone" represent the numbers of installed units.
*Compiled by TCA based on data publicized by the Ministry of Internal Affairs and Communications

2-1-2 Trends in Number of Subscriber Telephone Contracts by Prefecture

(Contracts)

Pref.	FY2017 Total	FY2018 Total	FY2019 Total	FY2020 Total	NTT(Re-entry) Total	Business	Residential
Hokkaido	986,548	923,739	851,620	796,415	769,563	137,075	632,488
Aomori	266,625	251,263	232,337	220,235	213,429	35,824	177,605
Iwate	245,938	233,019	216,909	206,255	199,472	33,526	165,946
Miyagi	339,839	318,343	296,178	279,251	265,162	54,962	210,200
Akita	199,855	188,956	176,344	167,366	161,846	28,100	133,746
Yamagata	171,957	160,955	149,321	140,238	135,348	25,499	109,849
Fukushima	328,538	307,809	285,623	269,270	261,769	47,864	213,905
Ibaraki	428,704	400,105	370,700	348,577	335,886	61,918	273,968
Tochigi	286,363	266,751	247,955	232,351	223,055	41,657	181,398
Gunma	294,422	276,539	258,205	242,358	233,174	40,831	192,343
Saitama	880,136	817,897	757,130	708,569	672,104	119,932	552,172
Chiba	768,715	715,804	663,591	621,850	590,005	114,876	475,129
Tokyo	2,008,796	1,876,185	1,746,802	1,632,327	1,492,760	485,652	1,007,108
Kanagawa	1,122,205	1,041,101	962,496	895,725	839,184	180,023	659,161
Niigata	358,366	335,803	311,268	291,960	280,754	55,349	225,405
Toyama	150,742	139,585	127,722	117,353	112,462	25,931	86,531
Ishikawa	168,099	159,298	149,183	140,245	134,283	29,581	104,702
Fukui	98,907	88,915	81,638	75,826	72,965	20,473	52,492
Yamanashi	145,796	134,501	123,877	115,143	111,788	23,062	88,726
Nagano	351,126	324,681	297,636	275,624	265,623	58,394	207,229
Gifu	286,642	265,742	245,433	227,804	218,562	51,223	167,339
Shizuoka	531,745	494,447	454,097	416,951	394,145	90,772	303,373
Aichi	871,708	809,403	745,776	690,630	646,020	166,202	479,818
Mie	263,444	245,304	223,625	205,111	197,762	45,135	152,627
Shiga	147,322	138,045	128,055	119,017	113,312	29,612	83,700
Kyoto	366,308	344,377	319,745	297,333	280,253	63,291	216,962
Osaka	1,169,017	1,093,866	1,007,276	933,172	856,663	232,537	624,126
Hyogo	597,351	559,365	518,001	481,673	454,296	113,726	340,570
Nara	175,261	164,482	152,252	140,712	133,025	27,036	105,989
Wakayama	157,279	148,574	137,894	128,224	124,276	25,704	98,572
Tottori	86,353	81,943	76,073	71,072	68,926	15,938	52,988
Shimane	139,870	134,306	125,435	115,811	113,602	22,659	90,943
Okayama	304,737	286,727	266,902	248,164	238,351	48,028	190,323
Hiroshima	472,271	446,484	416,457	389,825	374,005	75,249	298,756
Yamaguchi	287,885	272,802	254,499	237,910	232,624	37,169	195,455
Tokushima	121,631	113,946	104,816	96,540	93,522	20,957	72,565
Kagawa	149,222	139,600	128,440	118,793	112,483	24,549	87,934
Ehime	251,024	234,922	217,179	201,157	195,537	36,667	158,870
Kochi	150,803	141,651	130,410	121,011	118,189	23,443	94,746
Fukuoka	708,115	661,901	608,481	561,601	527,299	115,936	411,363
Saga	117,064	109,016	100,260	92,939	89,613	17,843	71,770
Nagasaki	273,987	256,654	237,908	220,404	214,126	39,677	174,449
Kumamoto	298,020	280,380	260,663	240,309	233,405	42,806	190,599
Oita	218,726	203,951	188,985	175,422	169,889	32,268	137,621
Miyazaki	189,497	175,738	160,800	148,004	143,954	25,856	118,098
Kagoshima	337,552	315,219	290,522	264,769	258,515	44,859	213,656
Okinawa	175,580	162,126	147,354	134,449	129,469	31,067	98,402
Total	**18,450,091**	**17,242,220**	**15,953,873**	**14,855,745**	**14,102,455**	**3,120,738**	**10,981,717**

*Compiled by TCA based on data publicized by the Ministry of Internal Affairs and Communications and other organizations

NTT Subscribers by Region (FY 2020)

0   1   2

(Unit: million subscribers)

Legend   ■ : Business use
         □ : Residential use

2-1-3　Trends in Number of ISDN Contracts by Prefecture

(Contracts)

Pref.	Basic Interface							Primary Rate Interface				
	FY2017	FY2018	FY2019	FY2020				FY2017	FY2018	FY2019	FY2020	
	Total	Total	Total	Total	NTT East ·West (Re-entry)			Total	Total	Total	Total	NTT East · West (Re-entry)
					Total	Business	Residential					
Hokkaido	125,374	116,055	106,018	96,904	76,624	67,509	9,115	808	787	744	648	361
Aomori	24,456	22,536	20,891	19,441	15,164	14,344	820	124	118	112	112	81
Iwate	25,928	24,172	22,302	20,915	16,056	15,070	986	100	97	88	83	54
Miyagi	52,458	48,994	45,352	41,782	29,359	27,728	1,631	435	449	431	420	158
Akita	19,806	18,294	16,895	15,922	12,668	11,891	777	99	97	93	90	71
Yamagata	21,274	19,657	18,122	16,569	12,982	12,213	769	96	93	93	90	62
Fukushima	36,088	33,518	31,015	28,605	22,477	20,783	1,694	134	125	118	105	65
Ibaraki	50,787	46,688	42,538	39,402	29,402	27,471	1,931	277	250	219	214	145
Tochigi	37,890	34,712	31,698	29,360	21,562	19,902	1,660	271	263	252	242	177
Gunma	37,445	34,139	31,425	29,164	21,350	19,586	1,764	277	228	229	221	134
Saitama	127,097	117,783	108,487	101,027	64,876	58,525	6,351	923	898	918	862	398
Chiba	108,923	100,981	92,803	85,720	58,570	53,945	4,625	1,066	1,028	945	865	461
Tokyo	510,787	476,007	440,386	400,743	251,876	233,306	18,570	16,485	15,873	15,248	14,562	6,289
Kanagawa	180,573	167,789	156,573	144,260	95,925	87,373	8,552	2,752	2,668	2,549	2,436	1,155
Niigata	45,173	41,720	38,380	35,639	26,390	24,865	1,525	167	160	148	141	81
Toyama	23,629	22,293	20,319	18,538	14,626	13,444	1,182	175	159	149	135	79
Ishikawa	26,104	24,520	22,508	20,698	16,165	14,777	1,388	208	186	180	175	82
Fukui	16,937	15,667	14,269	13,162	10,830	10,159	671	88	75	71	66	56
Yamanashi	16,822	15,409	14,316	13,181	10,640	9,693	947	80	78	75	70	55
Nagano	46,038	41,981	38,466	35,286	27,508	24,888	2,620	224	200	189	170	86
Gifu	42,317	39,703	36,506	33,747	26,974	24,443	2,531	203	197	166	162	104
Shizuoka	78,385	73,513	67,137	61,060	43,646	41,283	2,363	411	386	377	343	220
Aichi	163,259	152,646	140,621	129,553	90,071	83,811	6,260	1,459	1,371	1,342	1,226	698
Mie	38,356	36,363	33,548	31,212	25,635	23,490	2,145	170	170	155	143	104
Shiga	27,381	25,892	23,739	21,818	16,876	15,653	1,223	152	144	137	121	56
Kyoto	57,341	54,208	49,791	45,485	31,504	28,178	3,326	350	341	336	324	181
Osaka	248,123	232,199	214,062	197,113	121,148	111,771	9,377	4,285	3,972	3,847	3,765	1,837
Hyogo	94,127	88,503	82,250	76,196	55,639	51,462	4,177	796	779	760	743	399
Nara	22,217	20,836	19,194	17,713	12,857	11,034	1,823	105	95	90	89	61
Wakayama	17,437	16,323	15,010	13,803	11,233	10,263	970	65	63	69	64	49
Tottori	12,901	12,032	11,182	10,344	9,024	8,263	761	65	54	52	46	35
Shimane	16,170	15,405	14,423	13,431	11,936	10,941	995	135	128	122	117	62
Okayama	43,059	40,761	37,761	35,287	28,110	25,763	2,347	238	221	210	191	138
Hiroshima	65,442	63,269	58,886	54,292	41,912	38,552	3,360	383	365	340	320	198
Yamaguchi	30,008	28,633	26,520	24,268	19,972	18,171	1,801	135	128	131	103	68
Tokushima	15,194	14,429	13,383	12,251	10,193	9,358	835	79	71	59	57	39
Kagawa	22,657	21,397	19,519	18,086	13,842	13,087	755	153	148	143	130	75
Ehime	27,581	25,832	23,655	21,325	17,790	16,326	1,464	170	152	142	123	71
Kochi	15,776	14,955	13,962	12,947	11,304	10,535	769	80	74	73	68	57
Fukuoka	117,118	111,003	102,674	94,743	62,162	58,404	3,758	1,121	1,068	1,008	921	359
Saga	14,702	13,904	12,970	11,951	9,506	8,799	707	59	60	56	54	46
Nagasaki	26,487	25,234	23,388	21,362	17,311	16,149	1,162	158	152	151	141	74
Kumamoto	34,446	32,442	30,041	27,381	21,673	20,299	1,374	196	183	162	143	78
Oita	26,393	25,078	23,144	21,683	17,477	16,167	1,310	109	97	89	81	43
Miyazaki	21,200	20,156	18,461	16,878	13,683	12,723	960	121	118	105	108	73
Kagoshima	32,514	30,663	28,422	25,802	21,247	19,909	1,338	137	123	121	118	74
Okinawa	23,503	22,202	20,665	19,043	14,959	14,456	503	283	252	232	223	132
Nationwide	2,867,683	2,680,496	2,473,677	2,275,092	1,612,734	1,486,762	125,972	36,407	34,744	33,326	31,631	15,381

*Compiled by TCA based on data publicized by the Ministry of Internal Affairs and Communications and other organizations

2-1-4 Trends in Number of Mobile Phone and PHS Contracts by Prefecture

(Contracts)

Pref.	FY2017	FY2018	FY2019	FY2020
Hokkaido	5,843,959	5,895,707	5,819,753	5,975,105
Aomori	1,192,605	1,193,077	1,176,981	1,193,270
Iwate	1,167,778	1,168,610	1,150,198	1,171,489
Miyagi	2,737,821	2,680,955	2,795,336	2,957,708
Akita	923,138	918,106	899,429	908,889
Yamagata	1,038,527	1,039,742	1,024,110	1,041,223
Fukushima	1,875,172	1,868,427	1,838,020	1,859,929
Ibaraki	2,916,082	2,912,004	2,856,172	2,899,444
Tochigi	1,960,781	1,959,606	1,944,132	1,985,280
Gunma	2,001,265	2,020,847	1,981,904	2,028,492
Saitama	7,836,813	7,896,874	7,686,590	7,901,584
Chiba	6,643,408	6,654,827	6,544,681	6,761,478
Tokyo	48,432,052	53,622,797	60,034,916	62,247,537
Kanagawa	10,489,043	10,362,330	10,149,863	10,864,406
Niigata	2,185,331	2,171,151	2,133,268	2,164,965
Toyama	1,078,515	1,089,369	1,082,649	1,131,203
Ishikawa	1,196,666	1,190,816	1,179,718	1,208,789
Fukui	787,978	785,987	770,213	787,995
Yamanashi	856,877	852,212	830,699	841,432
Nagano	2,109,586	2,209,218	2,509,160	3,284,352
Gifu	2,041,326	2,029,266	1,990,436	2,092,344
Shizuoka	3,784,624	3,859,571	3,814,373	3,946,736
Aichi	8,911,004	9,617,688	9,871,726	10,383,697
Mie	1,832,030	1,821,398	1,781,566	1,832,072
Shiga	1,396,090	1,388,804	1,365,235	1,406,632
Kyoto	2,846,721	2,848,874	2,801,816	2,891,224
Osaka	11,415,942	11,562,119	11,585,950	12,229,891
Hyogo	5,787,715	5,672,086	5,531,958	5,726,188
Nara	1,346,126	1,341,371	1,321,433	1,367,343
Wakayama	954,555	943,434	920,099	929,237
Tottori	551,930	547,967	533,619	541,380
Shimane	674,419	670,166	657,315	668,920
Okayama	1,991,408	1,976,981	1,929,221	1,970,231
Hiroshima	3,341,318	3,355,221	3,373,136	3,550,125
Yamaguchi	1,408,333	1,399,108	1,383,085	1,416,291
Tokushima	732,379	730,036	717,519	730,836
Kagawa	1,097,300	1,046,049	1,020,433	1,034,491
Ehime	1,387,275	1,394,763	1,376,297	1,414,327
Kochi	706,034	699,776	685,580	695,020
Fukuoka	8,113,926	9,278,106	10,316,489	11,669,800
Saga	809,666	804,274	787,075	809,684
Nagasaki	1,341,208	1,331,605	1,301,392	1,333,284
Kumamoto	1,791,146	1,787,918	1,755,511	1,837,404
Oita	1,137,315	1,147,839	1,135,313	1,151,247
Miyazaki	1,061,712	1,057,817	1,042,396	1,062,780
Kagoshima	1,584,914	1,577,438	1,545,044	1,568,619
Okinawa	1,470,177	1,490,457	1,562,300	1,580,520
Total	172,789,990	179,872,794	186,514,109	195,054,893

*Compiled by TCA based on data publicized by the Ministry of Internal Affairs and Communications

2-1-5 Trends in Number of Domestic Leased Circuits

(Thousand circuits)

	FY2016	FY2017	FY2018	FY2019	FY2020
General Leased Circuits (Frequency Band Use)	209	203	197	192	191
General Leased Circuits (Code Transmission)	21	20	19	18	17
High-Speed Digital Transmission Services	121	109	78	43	42

*Compiled by TCA based on data publicized by the Ministry of Internal Affairs and Communications

2-1-6 Trends in Number of Broadband Service Contracts, etc.

(Contracts)

		FY2018	FY2019	FY2020	FY2021
Internet connection service (for fixed communication)	(total of 55 providers)	41,271,495	41,882,985	42,743,542	43,165,134
Internet connection service (for mobile communication)	(total of 29 providers)	181,674,422	185,242,351	191,334,287	196,516,577
FTTH access service	(total of 305 providers)	31,668,714	33,084,964	35,016,693	36,669,874
DSL access service	(total of 13 providers)	1,729,646	1,397,840	1,073,135	689,816
CATV access service	(total of 234 providers)	6,836,853	6,712,063	6,584,060	6,469,642
FWA access service	(total of 27 providers)	4,576	4,343	3,549	3,111
BWA access service	(total of 95 providers)	66,240,686	71,200,105	75,703,994	79,709,876
3.9-4G mobile phone terminals packet communications service	(total of 5 providers)	136,642,057	152,623,405	154,366,473	139,054,534
5G mobile phone terminals packet communications service	(total of 5 providers)	—	24,040	14,185,509	45,018,488
Local 5G service	(total of 4 providers)	—	—	17	70
Mobile Phone and PHS terminal Internet connection service	(total of 5 providers)	179,617,886	186,310,026	194,935,826	203,269,615
Public radio LAN access service	(total of 21 providers)	113,352,609	119,071,867	125,051,323	101,005,848
IP-VPN service	(total of 45 providers)	616,892	659,271	660,031	660,213
Wide-area Ethernet service	(total of 81 providers)	622,370	643,819	662,524	678,402

*Compiled by TCA based on data publicized by the Ministry of Internal Affairs and Communications

2-2 Situation of Traffic

2-2-1 Situation of Total Traffic

2-2-1-1 Trends in Total Number of Calls

(100 Million calls)

Outgoing \ Incoming	Subscriber Telephoe/ISDN					IP Phone				
	FY2016	FY2017	FY2018	FY2019	FY2020	FY2016	FY2017	FY2018	FY2019	FY2020
Subscriber Telephone	91.2	76.9	65.8	53.8	42.3					
Public Telephone	0.8	0.7	0.6	0.5	0.4	1.6	1.4	1.3	1.2	1.2
ISDN	78.8	72.9	63.8	57.3	47.3					
IP Phone	115.7	120.2	121.5	121.1	110.2	11.2	11.5	12.1	12.0	11.3
Mobile phone/PHS	60.8	56.6	50.5	45.6	39.6	64.7	70.5	72.0	72.3	69.9
Total	**347.4**	**327.3**	**302.2**	**278.2**	**239.7**	**77.5**	**83.4**	**85.4**	**85.5**	**82.4**

Outgoing \ Incoming	Mobile phone/PHS					Total				
	FY2016	FY2017	FY2018	FY2019	FY2020	FY2016	FY2017	FY2018	FY2019	FY2020
Subscriber Telephone										
Public Telephone	25.6	23.0	21.2	19.5	17.4	198.1	174.9	152.7	132.2	108.6
ISDN										
IP Phone	27.8	29.2	30.4	31.3	32.1	154.8	160.9	164.0	164.3	153.5
Mobile phone/PHS	378.5	358.9	343.8	327.4	307.1	503.9	486.1	466.3	445.3	416.5
Total	**431.9**	**411.1**	**395.5**	**378.1**	**356.5**	**856.8**	**821.8**	**783.0**	**741.8**	**678.7**

*Compiled by TCA based on data publicized by the Ministry of Internal Affairs and Communications

2-2-1-2 Trends in Total Number of Calls Between Fixed Telephone and Mobile Telephone

(100 Million calls)

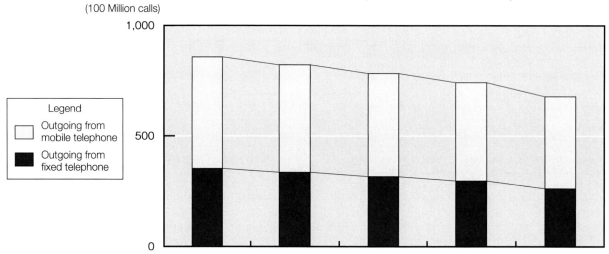

(100 Million calls)

Outgoing	Incoming	FY2016	FY2017	FY2018	FY2019	FY2020
Fixed	Fixed	299.3	283.6	265.1	245.9	212.7
Fixed	Mobile	53.4	52.2	51.6	50.8	49.5
Mobile	Mobile	378.5	358.9	343.8	327.4	307.1
Mobile	Fixed	125.5	127.1	122.5	117.9	109.5
Total		**856.8**	**821.8**	**783.0**	**741.8**	**678.7**

Note: Outgoing from fixed telephone: Outgoing from subscriber telephones, public telephones, ISDN and IP phones
Outgoing from mobile telephone: Outgoing from mobile phones and PHS
Incoming to fixed telephone: Incoming to subscriber telephones, ISDN and IP phones
Incoming to mobile telephone: Incoming to mobile phones and PHS

*Compiled by TCA based on data publicized by the Ministry of Internal Affairs and Communications

2-2-1-3 Trends in Daily Number of Calls per Subscription (Contract)

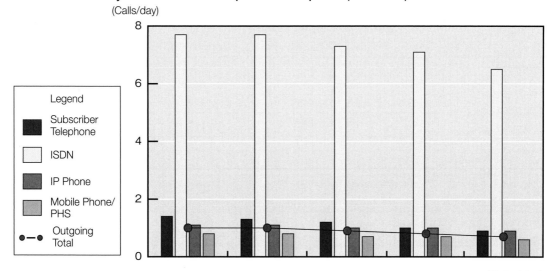

(Calls/day)

(Calls / day)

Outgoing	FY2016	FY2017	FY2018	FY2019	FY2020
Subscriber Telephone	1.4	1.3	1.2	1.0	0.9
ISDN	7.7	7.7	7.3	7.1	6.5
IP Phone	1.1	1.1	1.0	1.0	0.9
Mobile phone/PHS	0.8	0.8	0.7	0.7	0.6
Outgoing Total	**1.0**	**1.0**	**0.9**	**0.8**	**0.7**

Note: The categories of respective outgoing calls are as listed below. For example, the number of outgoing calls from subscriber telephones shows the total number of calls outgoing from subscriber telephones and destined for fixed telephones, IP phones, mobile phones, and PHS terminals. Since the actual number of outgoing calls from fixed telephones and destined for IP phones, mobile phones and PHS terminals cannot be identified, the number of those calls is calculated according to the ratio to the number of outgoing calls from fixed telephones and destined for fixed telephones.

Outgoing	ISDN	Cellular Telephone	PHS
Incoming	Fixed Telephone, IP Phone, Mobile Phone, PHS	Fixed Telephone, IP Phone, Mobile Phone, PHS	Fixed Telephone, IP Phone, Mobile Phone, PHS

*Compiled by TCA based on data publicized by the Ministry of Internal Affairs and Communications

2-2-1-4 Trends in Total Call Duration

(Million hours)

Incoming / Outgoing	Subscriber Telephoe/ISDN					IP Phone				
	FY2016	FY2017	FY2018	FY2019	FY2020	FY2016	FY2017	FY2018	FY2019	FY2020
Subscriber Telephone	288.0	234.3	194.6	154.3	130.1	5.8	5.0	4.4	4.2	4.3
Public Telephone	1.8	1.5	1.3	1.1	1.0					
ISDN	186.2	169.6	153.3	138.4	115.2					
IP Phone	359.5	351.7	340.4	327.5	304.2	49.9	48.3	49.9	48.2	48.7
Mobile phone/PHS	210.2	201.5	194.6	183.9	183.9	220.9	256.3	276.5	303.2	334.1
Total	**1,045.7**	**958.6**	**884.1**	**805.2**	**734.3**	**276.7**	**309.6**	**330.8**	**355.6**	**387.1**

Incoming / Outgoing	Mobile phone/PHS					Total				
	FY2016	FY2017	FY2018	FY2019	FY2020	FY2016	FY2017	FY2018	FY2019	FY2020
Subscriber Telephone	74.0	67.6	63.3	59.3	60.3	555.8	478.0	416.9	357.3	310.9
Public Telephone										
ISDN										
IP Phone	83.8	89.3	93.6	97.8	114.1	493.3	489.2	483.9	473.5	466.9
Mobile phone/PHS	1,800.2	1,722.6	1,656.1	1,607.1	1,736.2	2,231.4	2,180.4	2,127.2	2,094.2	2,254.2
Total	**1,958.1**	**1,879.4**	**1,813.0**	**1,764.2**	**1,910.6**	**3,280.5**	**3,147.6**	**3,027.9**	**2,925.0**	**3,032.1**

*Compiled by TCA based on data publicized by the Ministry of Internal Affairs and Communications

2-2-1-5 Trends in Average Call Duration per Call

(Seconds)

Outgoing \ Incoming	Subscriber Telephoe/ISDN					IP Phone				
	FY2016	FY2017	FY2018	FY2019	FY2020	FY2016	FY2017	FY2018	FY2019	FY2020
Subscriber Telephone	113.7	109.7	106.5	103.2	110.7	130.5	128.6	121.8	126.0	129.0
Public Telephone	81.0	77.1	78.0	79.2	90.0					
ISDN	85.1	83.8	86.5	87.0	87.7					
IP Phone	111.9	105.3	100.9	97.4	99.4	160.4	151.2	148.5	144.6	155.2
Mobile phone/PHS	124.5	128.2	138.7	145.2	167.2	122.9	130.9	138.3	151.0	172.1
Total	**108.4**	**105.4**	**105.3**	**104.2**	**110.3**	**128.5**	**133.6**	**139.4**	**149.7**	**169.1**

Outgoing \ Incoming	Mobile phone/PHS					Total				
	FY2016	FY2017	FY2018	FY2019	FY2020	FY2016	FY2017	FY2018	FY2019	FY2020
Subscriber Telephone										
Public Telephone	104.1	105.8	107.5	109.5	124.8	101.0	98.4	98.3	97.3	103.1
ISDN										
IP Phone	108.5	110.1	110.8	112.5	128.0	114.7	109.5	106.2	103.7	109.5
Mobile phone/PHS	171.2	172.8	173.4	176.7	203.5	159.4	161.5	164.2	169.3	194.8
Total	**163.2**	**164.6**	**165.0**	**168.0**	**192.9**	**137.8**	**137.9**	**139.2**	**142.0**	**160.8**

Note: Total Call Duration (seconds) ÷ Total Number of Calls (calls)
*Compiled by TCA based on data publicized by the Ministry of Internal Affairs and Communications

2-2-1-6 Trends in Daily Call Duration per Subscription (Contract)

(Seconds/day)

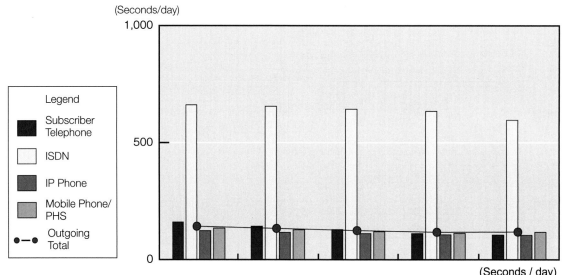

(Seconds / day)

Outgoing	FY2016	FY2017	FY2018	FY2019	FY2020
Subscriber Telephone	160	142	128	111	105
ISDN	661	655	643	634	596
IP Phone	123	116	111	106	104
Mobile phone/PHS	134	127	119	112	117
Outgoing Total	**142**	**133**	**124**	**117**	**118**

Note: The category of outgoing call duration and calculation method are the same as those in note of 2-2-1-3.
*Compiled by TCA based on data publicized by the Ministry of Internal Affairs and Communications

2-2-2 Situation of Traffic of Subscriber Telephone/ISDN

2-2-2-1 Situation of Calls by Time Zone

2-2-2-1-1 Trends in Number of Calls by Time Zone

(Million calls)

Time Zone	FY2016	FY2017	FY2018	FY2019	FY2020
0-1	129	116	100	87	71
1-2	113	102	89	79	66
2-3	101	92	81	71	61
3-4	94	84	75	67	58
4-5	96	86	76	68	60
5-6	118	107	93	81	73
6-7	166	148	130	113	97
7-8	321	283	244	202	164
8-9	789	697	616	509	415
9-10	1,645	1,454	1,267	1,085	869
10-11	1,710	1,518	1,323	1,132	919
11-12	1,586	1,406	1,227	1,055	862
12-13	963	852	733	626	519
13-14	1,401	1,242	1,074	925	757
14-15	1,403	1,247	1,082	932	767
15-16	1,403	1,245	1,077	933	768
16-17	1,413	1,244	1,083	939	760
17-18	1,203	1,048	905	774	597
18-19	834	714	602	503	381
19-20	584	493	410	344	260
20-21	385	322	267	226	175
21-22	238	201	169	144	109
22-23	171	149	128	109	82
23-24	145	128	111	95	74
Total	**17,003**	**14,975**	**12,961**	**11,103**	**8,966**

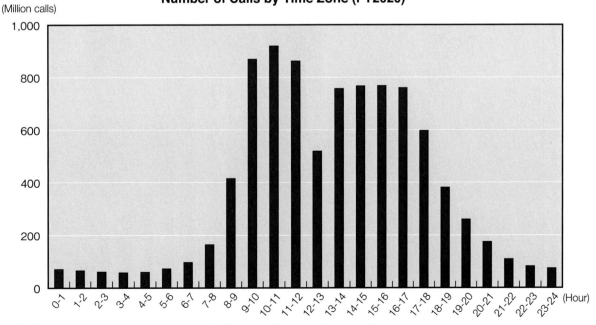

Number of Calls by Time Zone (FY2020)

(Million calls)

*Compiled by TCA based on data publicized by the Ministry of Internal Affairs and Communications

2-2-2-1-2   Trends in Call Duration by Time Zone

(Million hours)

Time Zone	FY2016	FY2017	FY2018	FY2019	FY2020
0-1	1.93	1.59	1.27	1.07	0.78
1-2	1.44	1.20	1.01	0.87	0.67
2-3	1.17	1.00	0.84	0.74	0.57
3-4	1.52	1.26	1.13	1.01	0.79
4-5	1.23	1.03	0.90	1.24	0.68
5-6	1.37	1.20	1.00	0.86	0.71
6-7	2.26	1.91	1.68	1.43	1.12
7-8	5.93	5.06	4.26	3.44	2.66
8-9	18.70	15.94	13.87	11.37	9.29
9-10	46.43	39.72	34.44	29.05	24.05
10-11	48.06	41.54	36.28	30.71	26.48
11-12	42.56	36.85	32.32	27.51	23.81
12-13	26.26	22.58	19.66	16.53	14.53
13-14	39.46	34.16	29.78	25.39	22.25
14-15	40.08	34.78	30.35	26.02	22.97
15-16	41.10	35.66	31.15	26.68	23.35
16-17	42.58	36.74	32.02	27.38	23.19
17-18	35.16	29.75	25.50	21.24	16.79
18-19	25.57	21.15	17.60	14.09	10.81
19-20	21.23	17.15	13.99	11.08	8.64
20-21	15.82	12.42	9.98	7.85	6.05
21-22	8.07	6.21	4.96	3.90	2.88
22-23	3.84	3.00	2.40	1.94	1.31
23-24	2.41	1.93	1.56	1.27	0.90
Total	**474.17**	**403.85**	**347.90**	**292.71**	**245.27**

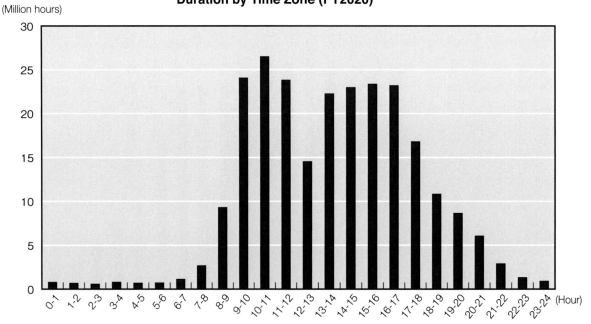

Duration by Time Zone (FY2020)

(Million hours)

*Compiled by TCA based on data publicized by the Ministry of Internal Affairs and Communications

2-2-2-2 Situation of Number of Calls by Duration

2-2-2-2-1 Trends in Number of Calls by Duration

(Million calls)

Duration	FY2016	FY2017	FY2018	FY2019	FY2020
up to 1 min	11,297	10,064	8,709	7,515	6,122
1-3 mins	3,696	3,217	2,798	2,364	1,828
over 3 mins	2,008	1,693	1,454	1,225	1,019
Total	**17,003**	**14,975**	**12,961**	**11,103**	**8,966**

Number of Calls by Duration

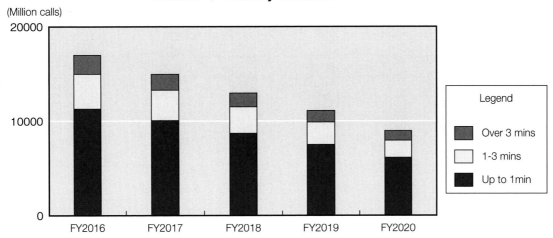

*Compiled by TCA based on data publicized by the Ministry of Internal Affairs and Communications

2-2-2-2-2 Number of Calls by Duration (10-second steps) (FY2020)

(Million calls)

Step	Total
～10 sec.	1,018
～20 sec.	1,307
～30 sec.	1,568
～40 sec.	1,100
～50 sec.	665
～60 sec.	464
～70 sec.	351
～80 sec.	266
～90 sec.	215
～100 sec.	183
～110 sec.	152
～120 sec.	133
～130 sec.	121
～140 sec.	102
～150 sec.	89
～160 sec.	79
～170 sec.	70
～180 sec.	67
180 sec. ～	1,019
Total	**8,966**

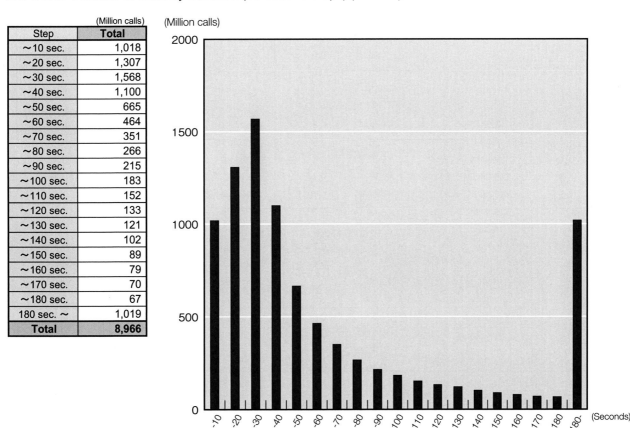

*Compiled by TCA based on data publicized by the Ministry of Internal Affairs and Communications

2-2-2-3 Situation of Calls by Prefecture

2-2-2-3-1   Ranking of Number of Outgoing and Incoming Calls by Prefecture (FY2020)

(Million calls)

Ranking	Outgoing			Incoming		
	Pref.	Number of outgoing calls	Ratio (%)	Pref.	Number of incoming calls	Ratio (%)
1	Tokyo	1,773	19.8	Tokyo	1,639	18.3
2	Osaka	883	9.8	Osaka	874	9.7
3	Kanagawa	577	6.4	Kanagawa	530	5.9
4	Aichi	485	5.4	Aichi	514	5.7
5	Saitama	462	5.2	Saitama	396	4.4
6	Hokkaido	379	4.2	Fukuoka	376	4.2
7	Fukuoka	371	4.1	Hokkaido	370	4.1
8	Hyogo	335	3.7	Chiba	357	4.0
9	Chiba	327	3.6	Hyogo	303	3.4
10	Shizuoka	226	2.5	Shizuoka	239	2.7
11	Hiroshima	192	2.1	Hiroshima	212	2.4
12	Miyagi	171	1.9	Miyagi	193	2.1
13	Kyoto	158	1.8	Kyoto	189	2.1
14	Niigata	145	1.6	Niigata	168	1.9
15	Ibaraki	141	1.6	Ibaraki	143	1.6
16	Nagano	124	1.4	Nagano	142	1.6
17	Gifu	116	1.3	Gifu	124	1.4
18	Okayama	115	1.3	Okayama	117	1.3
19	Fukushima	112	1.2	Gunma	117	1.3
20	Kagoshima	107	1.2	Fukushima	115	1.3
21	Gunma	107	1.2	Tochigi	105	1.2
22	Kumamoto	96	1.1	Kumamoto	101	1.1
23	Tochigi	94	1.0	Kagoshima	101	1.1
24	Mie	94	1.0	Mie	100	1.1
25	Yamaguchi	83	0.9	Iwate	83	0.9
26	Iwate	82	0.9	Yamaguchi	81	0.9
27	Aomori	79	0.9	Nagasaki	79	0.9
28	Nagasaki	79	0.9	Aomori	78	0.9
29	Ehime	71	0.8	Ehime	78	0.9
30	Shiga	70	0.8	Ishikawa	75	0.8
31	Oita	69	0.8	Yamagata	73	0.8
32	Ishikawa	68	0.8	Okinawa	70	0.8
33	Yamagata	67	0.7	Oita	68	0.8
34	Kagawa	66	0.7	Kagawa	68	0.8
35	Okinawa	64	0.7	Shiga	68	0.8
36	Akita	63	0.7	Toyama	68	0.8
37	Miyazaki	61	0.7	Akita	66	0.7
38	Toyama	60	0.7	Miyazaki	64	0.7
39	Nara	59	0.7	Shimane	62	0.7
40	Shimane	53	0.6	Nara	59	0.7
41	Wakayama	53	0.6	Wakayama	55	0.6
42	Kochi	42	0.5	Fukui	45	0.5
43	Yamanashi	41	0.5	Saga	43	0.5
44	Fukui	40	0.4	Yamanashi	43	0.5
45	Saga	39	0.4	Kochi	43	0.5
46	Tokushima	36	0.4	Tokushima	38	0.4
47	Tottori	33	0.4	Tottori	35	0.4
	Total	**8,966**	**100.0**	**Total**	**8,966**	**100.0**

*Compiled by TCA based on data publicized by the Ministry of Internal Affairs and Communications

2-2-2-3-2 Main Destination Prefectures by Originating Prefecture (FY2020)

Outgoing	Total Number of Outgoing calls (million)	Incoming									
		1		2		3		4		5	
		Pref.	Ratio (%)	Pref.	Ratio (%)	Pref.	Ratio (%)	Pref.	Ratio (%)	Pref.	Ratio (%)
Hokkaido	379	Hokkaido	76.6	Tokyo	8.0	Miyagi	2.5	Kanagawa	1.4	Saitama	1.1
Aomori	79	Aomori	73.3	Miyagi	7.6	Tokyo	5.7	Iwate	2.7	Akita	1.5
Iwate	82	Iwate	71.1	Miyagi	10.1	Tokyo	5.7	Aomori	2.5	Yamagata	1.5
Miyagi	171	Miyagi	62.6	Tokyo	9.1	Fukushima	4.3	Iwate	3.4	Yamagata	3.1
Akita	63	Akita	74.5	Miyagi	6.6	Tokyo	5.7	Yamagata	2.1	Aomori	1.6
Yamagata	67	Yamagata	71.9	Miyagi	9.0	Tokyo	6.5	Kanagawa	1.2	Saitama	1.1
Fukushima	112	Fukushima	68.0	Miyagi	10.0	Tokyo	9.6	Saitama	1.4	Kanagawa	1.3
Ibaraki	141	Ibaraki	56.3	Tokyo	12.4	Chiba	8.0	Saitama	7.6	Tochigi	2.5
Tochigi	94	Tochigi	61.0	Tokyo	13.1	Saitama	7.1	Ibaraki	3.3	Gunma	2.9
Gunma	107	Gunma	57.8	Tokyo	13.1	Saitama	6.3	Niigata	4.3	Tochigi	3.4
Saitama	462	Saitama	43.4	Tokyo	18.9	Chiba	5.3	Kanagawa	3.3	Gunma	2.2
Chiba	327	Chiba	59.1	Tokyo	20.8	Saitama	3.9	Kanagawa	3.0	Ibaraki	2.0
Tokyo	1,773	Tokyo	54.0	Kanagawa	6.6	Saitama	5.2	Osaka	4.5	Chiba	3.9
Kanagawa	577	Kanagawa	53.3	Tokyo	21.3	Osaka	2.7	Saitama	2.5	Chiba	2.1
Niigata	145	Niigata	76.9	Tokyo	7.8	Saitama	1.5	Osaka	1.4	Kanagawa	1.3
Toyama	60	Toyama	68.8	Ishikawa	5.6	Tokyo	5.5	Osaka	4.4	Kyoto	3.0
Ishikawa	68	Ishikawa	62.6	Tokyo	6.2	Toyama	4.9	Osaka	4.8	Kyoto	3.8
Fukui	40	Fukui	70.3	Osaka	5.8	Tokyo	4.9	Ishikawa	4.3	Kyoto	3.5
Yamanashi	41	Yamanashi	61.7	Tokyo	13.5	Saitama	7.1	Shizuoka	4.8	Kanagawa	2.8
Nagano	124	Nagano	69.0	Tokyo	9.0	Chiba	4.3	Niigata	4.0	Aichi	2.2
Gifu	116	Gifu	62.4	Aichi	17.4	Tokyo	5.2	Osaka	3.8	Kanagawa	1.1
Shizuoka	226	Shizuoka	70.5	Tokyo	7.7	Aichi	7.5	Osaka	3.1	Kanagawa	2.9
Aichi	485	Aichi	66.4	Tokyo	6.9	Osaka	4.8	Hyogo	3.6	Gifu	3.1
Mie	94	Mie	66.0	Aichi	12.8	Osaka	5.4	Tokyo	5.2	Kanagawa	1.1
Shiga	70	Shiga	52.6	Osaka	16.6	Kyoto	11.8	Tokyo	5.0	Aichi	2.2
Kyoto	158	Kyoto	60.6	Osaka	15.7	Tokyo	5.7	Shiga	2.8	Hyogo	2.2
Osaka	883	Osaka	56.1	Tokyo	7.7	Hyogo	5.5	Aichi	3.5	Kyoto	3.0
Hyogo	335	Hyogo	49.1	Osaka	21.6	Tokyo	6.2	Kyoto	2.2	Aichi	1.7
Nara	59	Nara	52.7	Osaka	21.3	Kyoto	9.7	Tokyo	4.9	Hyogo	1.6
Wakayama	53	Wakayama	64.0	Osaka	13.2	Tokyo	5.7	Kyoto	4.5	Aichi	1.6
Tottori	33	Tottori	66.4	Shimane	9.1	Osaka	4.6	Hiroshima	4.6	Tokyo	4.3
Shimane	53	Shimane	62.4	Tokyo	9.3	Hiroshima	6.2	Osaka	5.5	Tottori	2.9
Okayama	115	Okayama	63.3	Hiroshima	9.3	Osaka	6.6	Tokyo	5.0	Hyogo	3.6
Hiroshima	192	Hiroshima	69.7	Osaka	5.3	Tokyo	4.8	Okayama	3.6	Yamaguchi	2.8
Yamaguchi	83	Yamaguchi	65.5	Fukuoka	9.5	Hiroshima	8.5	Tokyo	4.4	Osaka	3.9
Tokushima	36	Tokushima	68.8	Osaka	5.9	Kagawa	5.6	Tokyo	4.9	Hiroshima	3.7
Kagawa	66	Kagawa	61.7	Osaka	6.5	Tokyo	5.3	Ehime	4.6	Hiroshima	3.7
Ehime	71	Ehime	70.4	Osaka	5.3	Tokyo	5.2	Hiroshima	4.1	Kagawa	3.9
Kochi	42	Kochi	75.0	Tokyo	4.5	Osaka	4.4	Kagawa	3.4	Hiroshima	2.7
Fukuoka	371	Fukuoka	63.6	Tokyo	5.9	Osaka	5.1	Kumamoto	2.2	Saga	1.9
Saga	39	Saga	66.3	Fukuoka	16.6	Tokyo	4.0	Osaka	2.8	Nagasaki	2.7
Nagasaki	79	Nagasaki	71.4	Fukuoka	9.9	Tokyo	4.4	Osaka	2.9	Saga	1.6
Kumamoto	96	Kumamoto	69.8	Fukuoka	11.2	Tokyo	4.1	Osaka	3.1	Kagoshima	1.5
Oita	69	Oita	70.5	Fukuoka	11.2	Tokyo	3.7	Osaka	3.1	Hyogo	2.5
Miyazaki	61	Miyazaki	73.3	Fukuoka	6.6	Tokyo	4.0	Kagoshima	3.0	Osaka	2.8
Kagoshima	107	Kagoshima	69.0	Fukuoka	6.0	Tokyo	3.8	Osaka	3.3	Kumamoto	2.3
Okinawa	64	Okinawa	71.9	Tokyo	7.1	Osaka	5.4	Fukuoka	4.5	Kanagawa	1.3

*Compiled by TCA based on data publicized by the Ministry of Internal Affairs and Communications

2-2-2-3-3   Main Originating Prefectures by Destination Prefecture (FY2020)

Incoming	Total number of incoming calls (million)	Outgoing									
		1		2		3		4		5	
		Pref.	Ratio (%)	Pref.	Ratio (%)	Pref.	Ratio (%)	Pref.	Ratio (%)	Pref.	Ratio (%)
Hokkaido	370	Hokkaido	78.4	Tokyo	8.5	Saitama	2.7	Osaka	1.8	Kanagawa	1.5
Aomori	78	Aomori	74.1	Tokyo	7.1	Miyagi	3.9	Saitama	2.8	Iwate	2.6
Iwate	83	Iwate	69.7	Tokyo	7.7	Miyagi	6.9	Saitama	2.8	Aomori	2.6
Miyagi	193	Miyagi	55.6	Tokyo	9.7	Fukushima	5.8	Hokkaido	4.8	Iwate	4.3
Akita	66	Akita	71.3	Tokyo	7.7	Miyagi	4.0	Saitama	2.9	Iwate	1.9
Yamagata	73	Yamagata	65.5	Tokyo	8.2	Miyagi	7.2	Saitama	3.2	Kanagawa	1.9
Fukushima	115	Fukushima	66.3	Tokyo	9.7	Miyagi	6.5	Saitama	3.7	Kanagawa	1.8
Ibaraki	143	Ibaraki	55.4	Tokyo	15.2	Saitama	6.8	Chiba	4.6	Kanagawa	3.1
Tochigi	105	Tochigi	54.4	Tokyo	15.0	Saitama	7.2	Kanagawa	3.7	Gunma	3.5
Gunma	117	Gunma	52.8	Tokyo	14.2	Saitama	8.8	Kanagawa	3.5	Osaka	2.6
Saitama	396	Saitama	50.6	Tokyo	23.2	Kanagawa	3.6	Chiba	3.2	Ibaraki	2.7
Chiba	357	Chiba	54.1	Tokyo	19.5	Saitama	6.8	Kanagawa	3.4	Ibaraki	3.1
Tokyo	1,639	Tokyo	58.4	Kanagawa	7.5	Saitama	5.3	Osaka	4.2	Chiba	4.1
Kanagawa	530	Kanagawa	58.1	Tokyo	22.1	Saitama	2.9	Osaka	2.9	Chiba	1.9
Niigata	168	Niigata	66.5	Tokyo	9.2	Osaka	3.5	Saitama	3.3	Nagano	3.0
Toyama	68	Toyama	61.1	Tokyo	8.9	Osaka	6.1	Ishikawa	4.9	Kanagawa	2.9
Ishikawa	75	Ishikawa	56.5	Tokyo	7.8	Osaka	6.2	Toyama	4.5	Aichi	3.9
Fukui	45	Fukui	61.3	Tokyo	8.5	Osaka	7.7	Ishikawa	4.5	Hyogo	2.5
Yamanashi	43	Yamanashi	58.5	Tokyo	18.3	Kanagawa	5.2	Saitama	2.7	Osaka	2.5
Nagano	142	Nagano	60.4	Tokyo	10.7	Osaka	6.4	Saitama	3.6	Kanagawa	2.5
Gifu	124	Gifu	58.1	Aichi	12.1	Tokyo	7.9	Osaka	5.2	Saitama	2.7
Shizuoka	239	Shizuoka	66.7	Tokyo	10.9	Aichi	4.3	Osaka	4.1	Kanagawa	3.4
Aichi	514	Aichi	62.6	Tokyo	9.0	Osaka	6.0	Gifu	3.9	Shizuoka	3.3
Mie	100	Mie	61.9	Aichi	10.2	Tokyo	7.9	Osaka	6.5	Hyogo	2.3
Shiga	68	Shiga	54.4	Osaka	12.5	Tokyo	8.6	Kyoto	6.4	Hyogo	3.6
Kyoto	189	Kyoto	50.6	Osaka	14.2	Tokyo	6.9	Shiga	4.4	Hyogo	4.0
Osaka	874	Osaka	56.8	Tokyo	9.1	Hyogo	8.3	Kyoto	2.8	Aichi	2.7
Hyogo	303	Hyogo	54.3	Osaka	15.9	Tokyo	7.9	Aichi	5.8	Kanagawa	2.0
Nara	59	Nara	52.9	Osaka	18.6	Tokyo	9.5	Hyogo	4.7	Kanagawa	2.6
Wakayama	55	Wakayama	61.7	Osaka	15.5	Tokyo	7.4	Hyogo	3.2	Kanagawa	2.7
Tottori	35	Tottori	62.4	Tokyo	7.4	Osaka	5.8	Hyogo	4.7	Shimane	4.4
Shimane	62	Shimane	53.7	Tokyo	10.4	Osaka	6.1	Hiroshima	4.8	Tottori	4.8
Okayama	117	Okayama	62.5	Osaka	7.7	Tokyo	7.5	Hiroshima	6.0	Hyogo	3.5
Hiroshima	212	Hiroshima	63.1	Tokyo	6.9	Osaka	5.3	Okayama	5.0	Yamaguchi	3.3
Yamaguchi	81	Yamaguchi	66.8	Tokyo	6.6	Hiroshima	6.5	Fukuoka	4.6	Osaka	4.3
Tokushima	38	Tokushima	65.0	Tokyo	7.7	Osaka	6.6	Kagawa	5.2	Hyogo	3.2
Kagawa	68	Kagawa	60.0	Tokyo	7.1	Osaka	7.0	Ehime	4.1	Tokushima	3.0
Ehime	78	Ehime	64.3	Tokyo	8.4	Osaka	6.3	Kagawa	3.9	Kanagawa	2.6
Kochi	43	Kochi	73.0	Tokyo	6.0	Osaka	4.9	Kagawa	3.3	Hyogo	1.9
Fukuoka	376	Fukuoka	62.8	Tokyo	7.6	Osaka	4.7	Kumamoto	2.9	Saitama	2.1
Saga	43	Saga	59.6	Fukuoka	16.5	Tokyo	6.4	Osaka	3.6	Nagasaki	2.9
Nagasaki	79	Nagasaki	71.7	Fukuoka	7.7	Tokyo	7.1	Osaka	2.9	Saitama	1.9
Kumamoto	101	Kumamoto	65.9	Fukuoka	8.1	Tokyo	7.1	Osaka	3.4	Kagoshima	2.5
Oita	68	Oita	71.3	Fukuoka	9.2	Tokyo	6.2	Osaka	2.8	Saitama	1.9
Miyazaki	64	Miyazaki	69.8	Tokyo	7.2	Fukuoka	5.6	Kagoshima	3.8	Osaka	2.9
Kagoshima	101	Kagoshima	73.5	Tokyo	6.5	Fukuoka	4.9	Osaka	3.0	Saitama	1.9
Okinawa	70	Okinawa	65.7	Tokyo	12.5	Osaka	4.4	Fukuoka	3.0	Saitama	2.0

*Compiled by TCA based on data publicized by the Ministry of Internal Affairs and Communications

2-2-2-4 Situation of Share by Carrier in Calls between Prefectures

2-2-2-4-1 Trends in Ratio of Number of Calls by Carrier in Calls between Prefectures

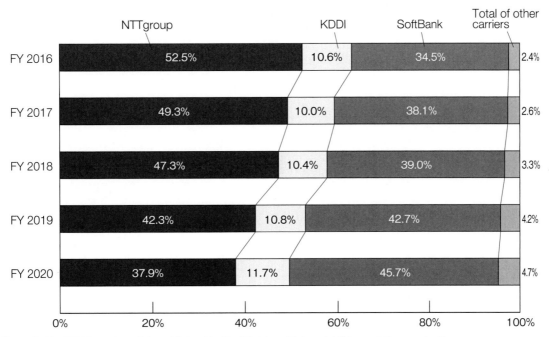

*Compiled by TCA based on data publicized by the Ministry of Internal Affairs and Communications

2-2-2-4-2 Trends in Ratio of Call Hours by Carrier in Calls between Prefectures

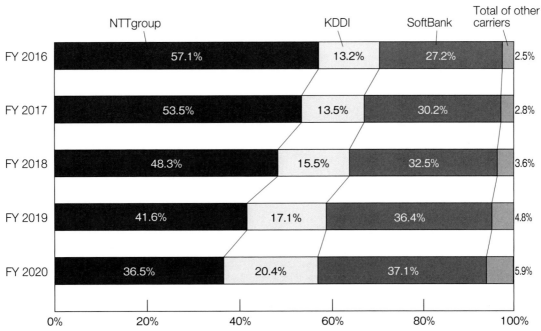

*Compiled by TCA based on data publicized by the Ministry of Internal Affairs and Communications

2-2-3 Situation of Traffic of IP Phones

2-2-3-1 Trends in Number of Telephone Numbers in Use and Communications Traffic

		FY2016		FY2017		FY2018		FY2019		FY2020	
Total number of numbers in use (million numbers)		40.99	(6.5%)	42.55	(3.8%)	43.41	(2.0%)	44.13	(1.7%)	44.67	(1.2%)
	(0ABJ-IP phone)	32.45	(5.5%)	33.64	(3.7%)	34.46	(2.4%)	35.21	(2.2%)	35.68	(1.3%)
	(050-IP phone)	8.53	(10.8%)	8.91	(4.5%)	8.95	(0.4%)	8.92	(▲0.3%)	8.99	(0.7%)
Number of calls (billion calls)		15.65	(3.3%)	16.23	(3.8%)	16.53	(1.8%)	16.55	(0.1%)	15.47	(▲6.5%)
	From IP phones to subscriber telephones, ISDN, IP phones, mobile phones, and PHS phones	15.49	(3.8%)	16.09	(3.9%)	16.40	(1.9%)	16.43	(0.2%)	15.35	(▲6.6%)
	From fixed-line services to IP phones	0.16	(▲29.7%)	0.14	(▲11.7%)	0.13	(▲11.0%)	0.12	(▲9.2%)	0.12	(2.4%)
Duration of calls (million hours)		499.3	(0.3%)	494.6	(▲1.0%)	488.5	(▲1.2%)	477.7	(▲2.2%)	471.2	(▲1.4%)
	From IP phones to subscriber telephones, ISDN, IP phones, mobile phones, and PHS phones	493.5	(1.0%)	489.5	(▲0.8%)	483.9	(▲1.1%)	473.5	(▲2.1%)	466.9	(▲1.4%)
	From fixed-line services to IP phones	5.8	(▲35.6%)	5.1	(▲12.7%)	4.7	(▲7.9%)	4.2	(▲10.0%)	4.3	(2.3%)

Notes: Figures in parentheses indicate rates of increase/decrease over the previous fiscal year.
*Compiled by TCA based on data publicized by the Ministry of Internal Affairs and Communications

2-2-4 Situation of Traffic of Mobile and PHS Phones

2-2-4-1 Situation of Calls by Time Zone

2-2-4-1-1 Trends in Number of Calls by Time Zone

(Calls to and from mobile / PHS phones)   (Million calls)

Time Zone	FY2016	FY2017	FY2018	FY2019	FY2020
0-1	358	318	276	248	176
1-2	229	208	181	164	118
2-3	161	152	132	121	90
3-4	128	125	110	100	78
4-5	126	126	112	104	86
5-6	206	204	186	174	149
6-7	520	503	470	440	373
7-8	1,223	1,188	1,136	1,073	929
8-9	2,423	2,373	2,317	2,222	2,021
9-10	3,754	3,696	3,638	3,530	3,394
10-11	4,021	3,952	3,877	3,768	3,728
11-12	3,915	3,828	3,739	3,627	3,609
12-13	3,439	3,306	3,170	3,031	2,881
13-14	3,662	3,567	3,474	3,355	3,311
14-15	3,591	3,505	3,420	3,315	3,299
15-16	3,891	3,802	3,706	3,582	3,524
16-17	4,258	4,150	4,036	3,889	3,761
17-18	4,693	4,515	4,328	4,118	3,820
18-19	4,042	3,818	3,586	3,351	2,969
19-20	3,010	2,798	2,586	2,393	2,044
20-21	2,179	2,000	1,824	1,670	1,375
21-22	1,512	1,360	1,224	1,107	857
22-23	962	854	753	679	497
23-24	585	515	447	400	289
Total	**52,889**	**50,864**	**48,728**	**46,460**	**43,379**

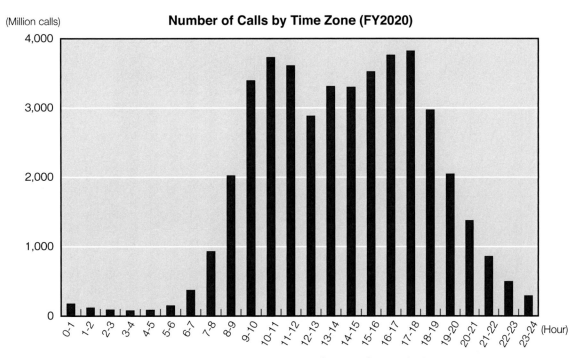

(Million calls)
Number of Calls by Time Zone (FY2020)

*Compiled by TCA based on data publicized by the Ministry of Internal Affairs and Communications

2-2-4-1-2 Trends in Call Duration by Time Zone

(Calls to and from mobile / PHS phones)   (Million hours)

Time Zone	FY2016	FY2017	FY2018	FY2019	FY2020
0-1	43.18	37.46	32.19	30.05	28.00
1-2	26.38	23.18	20.03	19.16	18.18
2-3	17.27	15.90	14.11	13.95	13.55
3-4	12.77	12.35	11.28	11.47	11.43
4-5	12.69	12.71	11.47	11.70	11.99
5-6	13.04	13.33	12.79	13.27	13.79
6-7	22.50	22.84	22.32	22.51	22.43
7-8	46.53	46.97	46.37	45.82	44.66
8-9	86.83	87.08	86.83	85.60	85.46
9-10	138.35	138.44	138.73	137.77	147.69
10-11	149.94	149.92	150.16	149.92	168.69
11-12	140.59	140.16	140.20	140.24	160.60
12-13	125.15	122.82	120.95	119.31	128.68
13-14	130.64	129.82	129.57	129.27	147.36
14-15	132.10	131.71	131.97	132.58	154.69
15-16	143.48	143.41	143.68	143.77	165.99
16-17	157.49	157.23	157.15	156.77	177.51
17-18	175.87	173.56	171.32	168.87	182.72
18-19	164.82	160.32	155.87	151.42	158.19
19-20	146.66	141.41	136.20	131.66	136.25
20-21	137.82	131.51	125.06	120.12	124.12
21-22	120.17	111.93	104.41	98.99	99.41
22-23	93.22	84.52	76.19	71.44	68.51
23-24	66.43	58.46	50.97	47.36	44.33
Total	**2,303.94**	**2,247.02**	**2,189.83**	**2,153.00**	**2,314.22**

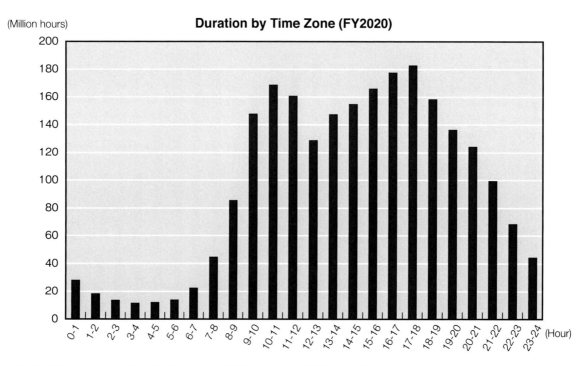

(Million hours)   **Duration by Time Zone (FY2020)**

*Compiled by TCA based on data publicized by the Ministry of Internal Affairs and Communications

2-2-4-2 Situation of Number of Calls by Duration

2-2-4-2-1 Trends in Number of Calls by Duration

(Calls to and from mobile / PHS phones) (Million calls)

Duration	FY2016	FY2017	FY2018	FY2019	FY2020
up to 1 min	29,177	27,701	26,235	24,894	22,107
1-3 mins	14,359	13,943	13,472	12,804	11,965
over 3 mins	9,352	9,219	9,020	8,763	9,309
Total	52,889	50,864	48,728	46,460	43,379

Number of Calls by Duration
(Calls to and from mobile / PHS phones)

(Million calls)

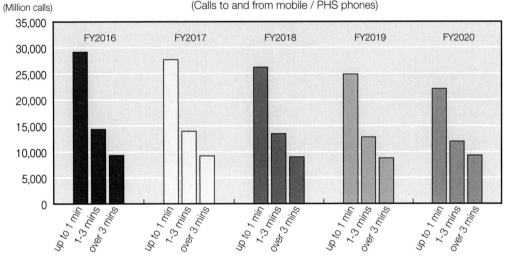

*Compiled by TCA based on data publicized by the Ministry of Internal Affairs and Communications

2-2-4-2-2 Number of Calls by Duration (10-second steps) (FY2020)

(Million calls)

Step	Calls to and from mobile/PHS phones
～10 sec.	4,991
～20 sec.	4,488
～30 sec.	4,176
～40 sec.	3,406
～50 sec.	2,766
～60 sec.	2,280
～70 sec.	1,911
～80 sec.	1,631
～90 sec.	1,398
～100sec.	1,213
～110sec.	1,061
～120sec.	933
～130sec.	826
～140sec.	733
～150sec.	656
～160sec.	589
～170sec.	531
～180sec.	483
180sec.～	9,309
Total	43,379

(Million calls)

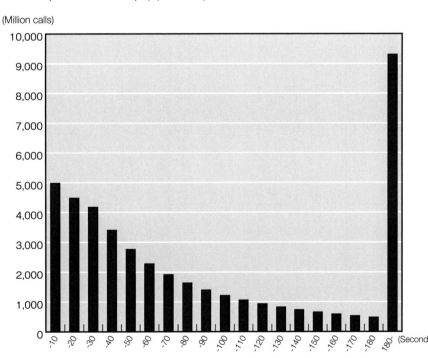

*Compiled by TCA based on data publicized by the Ministry of Internal Affairs and Communications

2-2-4-3 Situation of Calls by Prefecture

2-2-4-3-1 Ranking of Number of Outgoing and Incoming Calls by Prefecture (FY2020)

(Million calls)

Ranking	Outgoing			Incoming		
	Pref.	No. of Outgoing	Ratio (%)	Pref.	No. of Incoming	Ratio (%)
1	Tokyo	5,685	13.2	Tokyo	6,405	14.9
2	Osaka	3,418	7.9	Osaka	3,365	7.8
3	Kanagawa	2,528	5.9	Kanagawa	2,541	5.9
4	Aichi	2,442	5.7	Aichi	2,367	5.5
5	Fukuoka	2,107	4.9	Fukuoka	2,066	4.8
6	Saitama	2,043	4.7	Saitama	2,010	4.7
7	Chiba	1,844	4.3	Chiba	1,857	4.3
8	Hyogo	1,705	4.0	Hyogo	1,592	3.7
9	Hokkaido	1,593	3.7	Hokkaido	1,547	3.6
10	Shizuoka	1,153	2.7	Shizuoka	1,131	2.6
11	Hiroshima	993	2.3	Hiroshima	970	2.3
12	Ibaraki	975	2.3	Ibaraki	951	2.2
13	Kyoto	809	1.9	Kyoto	788	1.8
14	Miyagi	774	1.8	Miyagi	765	1.8
15	Kumamoto	712	1.7	Kumamoto	694	1.6
16	Okayama	706	1.6	Okayama	684	1.6
17	Kagoshima	679	1.6	Kagoshima	668	1.6
18	Okinawa	677	1.6	Okinawa	657	1.5
19	Tochigi	651	1.5	Nagano	647	1.5
20	Mie	649	1.5	Niigata	642	1.5
21	Niigata	648	1.5	Tochigi	640	1.5
22	Nagano	646	1.5	Mie	631	1.5
23	Fukushima	632	1.5	Fukushima	619	1.4
24	Gunma	625	1.5	Gunma	616	1.4
25	Gifu	624	1.5	Gifu	609	1.4
26	Ehime	496	1.2	Nagasaki	487	1.1
27	Nagasaki	495	1.2	Ehime	486	1.1
28	Oita	457	1.1	Oita	450	1.0
29	Yamaguchi	451	1.0	Yamaguchi	442	1.0
30	Shiga	433	1.0	Shiga	423	1.0
31	Miyazaki	420	1.0	Miyazaki	413	1.0
32	Nara	407	0.9	Nara	392	0.9
33	Kagawa	364	0.8	Kagawa	360	0.8
34	Wakayama	362	0.8	Ishikawa	359	0.8
35	Ishikawa	361	0.8	Wakayama	354	0.8
36	Iwate	337	0.8	Aomori	334	0.8
37	Aomori	337	0.8	Iwate	334	0.8
38	Yamagata	333	0.8	Yamagata	327	0.8
39	Saga	324	0.8	Yamanashi	310	0.7
40	Yamanashi	314	0.7	Saga	305	0.7
41	Toyama	303	0.7	Toyama	297	0.7
42	Kochi	283	0.7	Kochi	278	0.6
43	Akita	280	0.7	Akita	277	0.6
44	Tokushima	275	0.6	Tokushima	268	0.6
45	Fukui	269	0.6	Fukui	263	0.6
46	Shimane	217	0.5	Shimane	214	0.5
47	Tottori	185	0.4	Tottori	183	0.4
	Total	**43,018**	**100.0**	**Total**	**43,018**	**100.0**

Note: Compiled from data on calls to and from mobile and PHS phones.
*Compiled by TCA based on data publicized by the Ministry of Internal Affairs and Communications

2-2-4-3-2 Main Destination Prefectures by Originating Prefecture (FY2020)

Outgoing	Total number of outgoing calls (million)	Incoming									
		1		2		3		4		5	
		Pref.	Ratio (%)	Pref.	Ratio (%)	Pref.	Ratio (%)	Pref.	Ratio (%)	Pref.	Ratio (%)
Hokkaido	1,593	Hokkaido	90.3	Tokyo	4.9	Kanagawa	0.5	Saitama	0.4	Osaka	0.4
Aomori	337	Aomori	86.1	Tokyo	4.0	Miyagi	2.1	Iwate	2.0	Hokkaido	0.9
Iwate	337	Iwate	82.6	Miyagi	4.8	Tokyo	3.9	Aomori	2.1	Akita	1.2
Miyagi	774	Miyagi	80.6	Tokyo	5.6	Fukushima	2.5	Iwate	2.0	Yamagata	1.5
Akita	280	Akita	85.9	Tokyo	4.1	Miyagi	2.3	Iwate	1.4	Aomori	1.0
Yamagata	333	Yamagata	84.6	Tokyo	4.7	Miyagi	3.7	Fukushima	1.1	Kanagawa	0.7
Fukushima	632	Fukushima	82.7	Tokyo	5.4	Miyagi	3.5	Ibaraki	1.1	Saitama	0.9
Ibaraku	975	Ibaraki	77.8	Tokyo	7.3	Chiba	4.3	Saitama	2.5	Tochigi	2.3
Tochigi	651	Tochigi	78.0	Tokyo	6.3	Ibaraki	3.4	Saitama	3.0	Gunma	2.8
Gunma	625	Gunma	78.9	Tokyo	6.2	Saitama	5.1	Tochigi	3.0	Kanagawa	1.0
Saitama	2,043	Saitama	68.0	Tokyo	17.2	Chiba	2.8	Kanagawa	2.4	Gunma	1.6
Chiba	1,844	Chiba	72.8	Tokyo	14.0	Saitama	2.8	Kanagawa	2.3	Ibaraki	2.2
Tokyo	5,685	Tokyo	67.2	Kanagawa	7.3	Saitama	6.0	Chiba	4.6	Osaka	2.1
Kanagawa	2,528	Kanagawa	73.1	Tokyo	15.5	Chiba	1.7	Saitama	1.7	Shizuoka	1.0
Niigata	648	Niigata	86.9	Tokyo	4.8	Saitama	1.0	Nagano	0.8	Kanagawa	0.8
Toyama	303	Toyama	82.9	Ishikawa	4.4	Tokyo	4.4	Osaka	1.2	Aichi	1.1
Ishikawa	361	Ishikawa	82.3	Tokyo	4.4	Toyama	3.1	Fukui	1.9	Osaka	1.6
Fukui	269	Fukui	82.8	Tokyo	4.1	Ishikawa	3.0	Osaka	2.1	Aichi	1.2
Yamanashi	314	Yamanashi	82.4	Tokyo	7.3	Kanagawa	2.1	Nagano	1.5	Shizuoka	1.5
Nagano	646	Nagano	85.0	Tokyo	5.2	Aichi	1.3	Saitama	1.0	Kanagawa	0.9
Gifu	624	Gifu	76.6	Aichi	11.5	Tokyo	4.7	Osaka	1.0	Mie	0.9
Shizuoka	1,153	Shizuoka	83.1	Tokyo	6.0	Aichi	3.0	Kanagawa	2.0	Osaka	0.8
Aichi	2,442	Aichi	81.5	Tokyo	5.9	Gifu	2.9	Mie	1.6	Osaka	1.4
Mie	649	Mie	80.2	Aichi	6.4	Tokyo	4.4	Osaka	1.9	Gifu	0.9
Shiga	433	Shiga	74.6	Kyoto	6.0	Osaka	5.7	Tokyo	4.1	Aichi	1.3
Kyoto	809	Kyoto	72.4	Osaka	9.3	Tokyo	4.9	Shiga	3.2	Hyogo	2.3
Osaka	3,418	Osaka	75.6	Tokyo	6.9	Hyogo	4.6	Kyoto	2.1	Nara	1.6
Hyogo	1,705	Hyogo	73.9	Osaka	12.5	Tokyo	5.2	Kyoto	1.2	Okayama	0.6
Nara	407	Nara	70.3	Osaka	12.9	Tokyo	5.7	Kyoto	2.8	Hyogo	1.6
Wakayama	362	Wakayama	81.9	Osaka	7.9	Tokyo	3.7	Hyogo	1.0	Nara	1.0
Tottori	185	Tottori	81.6	Shimane	4.3	Tokyo	3.4	Okayama	1.9	Hiroshima	1.8
Shimane	217	Shimane	82.2	Hiroshima	3.8	Tottori	3.7	Tokyo	2.9	Osaka	1.4
Okayama	706	Okayama	82.3	Tokyo	4.2	Hiroshima	3.7	Osaka	1.9	Hyogo	1.5
Hiroshima	993	Hiroshima	82.3	Tokyo	4.6	Okayama	2.3	Yamaguchi	1.9	Osaka	1.6
Yamaguchi	451	Yamaguchi	81.7	Hiroshima	4.2	Tokyo	3.8	Fukuoka	3.7	Osaka	1.1
Tokushima	275	Tokushima	84.6	Tokyo	3.4	Kagawa	3.1	Osaka	1.9	Hyogo	1.3
Kagawa	364	Kagawa	80.9	Tokyo	4.4	Ehime	2.6	Tokushima	2.0	Osaka	2.0
Ehime	496	Ehime	84.7	Tokyo	4.1	Kagawa	2.3	Osaka	1.5	Hiroshima	1.3
Kochi	283	Kochi	87.2	Tokyo	3.2	Ehime	1.8	Kagawa	1.7	Osaka	1.4
Fukuoka	2,107	Fukuoka	82.0	Tokyo	4.9	Saga	1.9	Kumamoto	1.6	Oita	1.3
Saga	324	Saga	71.8	Fukuoka	15.5	Tokyo	3.8	Nagasaki	3.1	Kumamoto	1.0
Nagasaki	495	Nagasaki	84.9	Fukuoka	4.6	Tokyo	3.4	Saga	1.9	Kumamoto	0.8
Kumamoto	712	Kumamoto	83.9	Fukuoka	5.1	Tokyo	3.9	Kagoshima	1.2	Miyazaki	0.7
Oita	457	Oita	83.9	Fukuoka	5.9	Tokyo	3.6	Kumamoto	1.0	Osaka	0.7
Miyazaki	420	Miyazaki	84.9	Tokyo	3.7	Kagoshima	3.0	Fukuoka	2.6	Kumamoto	1.3
Kagoshima	679	Kagoshima	86.5	Tokyo	3.4	Fukuoka	2.4	Miyazaki	2.0	Kumamoto	1.2
Okinawa	677	Okinawa	90.7	Tokyo	4.4	Fukuoka	0.9	Osaka	0.7	Kanagawa	0.4

Note: Compiled from data on calls to and from mobile and PHS phones.
*Compiled by TCA based on data publicized by the Ministry of Internal Affairs and Communications

2-2-4-3-3　Main Originating Prefectures by Destination Prefecture (FY2020)

Incoming	Total number of Incoming calls (million)	Outgoing									
		1		2		3		4		5	
		Pref.	Ratio (%)	Pref.	Ratio (%)	Pref.	Ratio (%)	Pref.	Ratio (%)	Pref.	Ratio (%)
Hokkaido	1,547	Hokkaido	93.0	Tokyo	2.4	Kanagawa	0.5	Osaka	0.4	Saitama	0.4
Aomori	334	Aomori	86.7	Tokyo	2.7	Iwate	2.1	Miyagi	1.9	Akita	0.9
Iwate	334	Iwate	83.4	Miyagi	4.6	Tokyo	2.7	Aomori	2.1	Akita	1.2
Miyagi	765	Miyagi	81.5	Tokyo	3.6	Fukushima	2.9	Iwate	2.1	Yamagata	1.6
Akita	277	Akita	86.8	Tokyo	2.8	Miyagi	2.1	Iwate	1.4	Aomori	1.0
Yamagata	327	Yamagata	86.2	Miyagi	3.5	Tokyo	2.7	Fukushima	1.1	Kanagawa	0.8
Fukushima	619	Fukushima	84.4	Miyagi	3.2	Tokyo	3.2	Ibaraki	1.3	Saitama	1.0
Ibaraki	951	Ibaraki	79.7	Tokyo	5.5	Chiba	4.2	Saitama	2.5	Tochigi	2.4
Tochigi	640	Tochigi	79.2	Tokyo	4.6	Ibaraki	3.5	Saitama	3.3	Gunma	2.9
Gunma	616	Gunma	79.9	Saitama	5.4	Tokyo	4.6	Tochigi	2.9	Kanagawa	1.1
Saitama	2,010	Saitama	69.1	Tokyo	17.1	Chiba	2.6	Kanagawa	2.1	Gunma	1.6
Chiba	1,857	Chiba	72.3	Tokyo	13.9	Saitama	3.0	Kanagawa	2.4	Ibaraki	2.3
Tokyo	6,405	Tokyo	59.6	Kanagawa	6.1	Saitama	5.5	Chiba	4.0	Osaka	3.7
Kanagawa	2,541	Kanagawa	72.8	Tokyo	16.3	Saitama	1.9	Chiba	1.6	Shizuoka	0.9
Niigata	642	Niigata	87.7	Tokyo	3.4	Saitama	1.1	Kanagawa	0.9	Nagano	0.8
Toyama	297	Toyama	84.4	Ishikawa	3.8	Tokyo	2.9	Aichi	1.2	Osaka	1.0
Ishikawa	359	Ishikawa	82.7	Toyama	3.7	Tokyo	2.6	Fukui	2.2	Osaka	1.5
Fukui	263	Fukui	84.9	Ishikawa	2.6	Tokyo	2.2	Osaka	1.8	Aichi	1.3
Yamanashi	310	Yamanashi	83.4	Tokyo	6.2	Kanagawa	2.2	Nagano	1.5	Shizuoka	1.5
Nagano	647	Nagano	84.9	Tokyo	4.2	Saitama	1.4	Aichi	1.3	Kanagawa	1.1
Gifu	609	Gifu	78.5	Aichi	11.7	Tokyo	2.3	Osaka	1.1	Mie	0.9
Shizuoka	1,131	Shizuoka	84.7	Tokyo	4.1	Aichi	2.9	Kanagawa	2.2	Osaka	0.9
Aichi	2,367	Aichi	84.0	Gifu	3.0	Tokyo	2.9	Mie	1.8	Shizuoka	1.5
Mie	631	Mie	82.5	Aichi	6.3	Osaka	2.2	Tokyo	1.9	Gifu	0.9
Shiga	423	Shiga	76.5	Kyoto	6.1	Osaka	5.9	Tokyo	1.9	Hyogo	1.4
Kyoto	788	Kyoto	74.3	Osaka	9.2	Shiga	3.3	Tokyo	2.6	Hyogo	2.5
Osaka	3,365	Osaka	76.8	Hyogo	6.3	Tokyo	3.5	Kyoto	2.2	Nara	1.6
Hyogo	1,592	Hyogo	79.2	Osaka	9.9	Tokyo	2.4	Kyoto	1.2	Okayama	0.7
Nara	392	Nara	72.9	Osaka	13.6	Kyoto	3.0	Tokyo	1.9	Hyogo	1.7
Wakayama	354	Wakayama	83.6	Osaka	7.9	Tokyo	1.5	Hyogo	1.2	Nara	1.1
Tottori	183	Tottori	82.6	Shimane	4.4	Okayama	2.0	Tokyo	2.0	Osaka	1.9
Shimane	214	Shimane	83.4	Tottori	3.8	Hiroshima	3.4	Tokyo	1.8	Osaka	1.3
Okayama	684	Okayama	85.0	Hiroshima	3.4	Osaka	1.9	Tokyo	1.8	Hyogo	1.5
Hiroshima	970	Hiroshima	84.3	Okayama	2.7	Tokyo	2.0	Yamaguchi	2.0	Osaka	1.5
Yamaguchi	442	Yamaguchi	83.3	Hiroshima	4.2	Fukuoka	3.8	Tokyo	1.8	Osaka	1.2
Tokushima	268	Tokushima	86.8	Kagawa	2.8	Osaka	1.8	Tokyo	1.6	Hyogo	1.4
Kagawa	360	Kagawa	81.7	Ehime	3.1	Tokushima	2.3	Tokyo	2.1	Osaka	1.9
Ehime	486	Ehime	86.5	Tokyo	2.4	Kagawa	1.9	Osaka	1.5	Hiroshima	1.3
Kochi	278	Kochi	88.9	Ehime	1.8	Tokyo	1.6	Osaka	1.4	Kagawa	1.4
Fukuoka	2,066	Fukuoka	83.7	Saga	2.4	Tokyo	2.4	Kumamoto	1.7	Oita	1.3
Saga	305	Saga	76.1	Fukuoka	13.1	Nagasaki	3.1	Tokyo	1.7	Kumamoto	1.0
Nagasaki	487	Nagasaki	86.3	Fukuoka	4.5	Saga	2.0	Tokyo	1.7	Kumamoto	0.8
Kumamoto	694	Kumamoto	86.1	Fukuoka	4.9	Tokyo	1.7	Kagoshima	1.2	Miyazaki	0.8
Oita	450	Oita	85.3	Fukuoka	6.2	Tokyo	1.6	Kumamoto	1.1	Miyazaki	0.7
Miyazaki	413	Miyazaki	86.2	Kagoshima	3.4	Fukuoka	2.4	Tokyo	1.8	Kumamoto	1.3
Kagoshima	668	Kagoshima	87.9	Fukuoka	2.2	Miyazaki	1.9	Tokyo	1.8	Kumamoto	1.2
Okinawa	657	Okinawa	93.5	Tokyo	2.0	Fukuoka	0.8	Osaka	0.6	Kanagawa	0.4

Note: Compiled from data on calls to and from mobile and PHS phones.
*Compiled by TCA based on data publicized by the Ministry of Internal Affairs and Communications

2-2-5 Situation of Traffic of International Telephone Services

2-2-5-1 Trends in Number and Duration of International Telephone Calls

(Million calls, Million minutes)

Category		FY2016	FY2017	FY2018	FY2019	FY2020
Number of Calls	Outgoing	212.8	194.8	159.1	137.9	50.0
	Incoming	259.5	298.6	289.3	333.5	317.6
	Total	**472.2**	**493.4**	**448.5**	**471.4**	**367.6**
Duration of calls	Outgoing	855.6	744.4	594.3	496.5	258.5
	Incoming	822.2	902.1	750.9	661.1	527.1
	Total	**1,677.8**	**1,646.5**	**1,345.2**	**1,157.6**	**785.7**

*Compiled by TCA based on data publicized by the Ministry of Internal Affairs and Communications

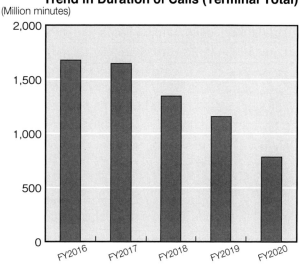

Trend in Number of Calls (Terminal Total)
(Million calls)

Trend in Duration of Calls (Terminal Total)
(Million minutes)

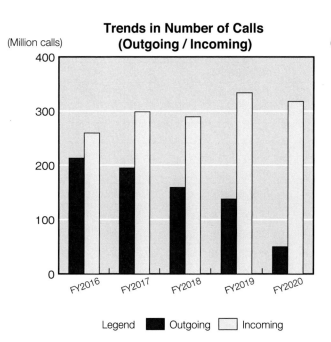

Trends in Number of Calls (Outgoing / Incoming)
(Million calls)

Legend ■ Outgoing □ Incoming

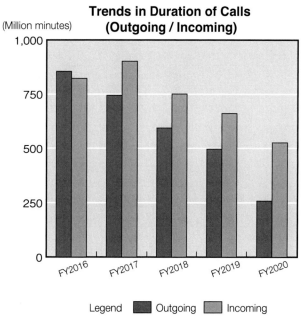

Trends in Duration of Calls (Outgoing / Incoming)
(Million minutes)

Legend ■ Outgoing ■ Incoming

2-2-5-2 Situation of Duration of International Calls by Country/Region (Top Countries/Regions Shown)

2-2-5-2-1 Trends in Share of Outgoing Call Duration by Country

Ranking	FY2016		FY2017		FY2018		FY2019		FY2020	
1	China	22.58%	China	20.93%	U.S.A. (mainland)	19.33%	U.S.A. (mainland)	19.83%	U.S.A. (mainland)	35.13%
2	U.S.A. (mainland)	14.52%	U.S.A. (mainland)	17.79%	China	17.75%	Hong Kong	19.19%	China	16.15%
3	Philippines	12.31%	Hong Kong	10.80%	Hong Kong	15.84%	China	16.46%	Hong Kong	8.86%
4	Korea	6.99%	Philippines	8.46%	Philippines	6.36%	Korea	5.16%	Korea	6.26%
5	Hong Kong	4.53%	Korea	6.01%	Korea	6.06%	Thailand	3.49%	Thailand	3.51%
6	Thailand	4.06%	Thailand	3.63%	Thailand	3.74%	Philippines	3.34%	Philippines	3.49%
7	Taiwan	3.45%	Taiwan	3.11%	Taiwan	3.19%	Taiwan	3.02%	Taiwan	3.20%
8	Singapore	2.50%	Singapore	2.83%	Singapore	2.80%	Singapore	2.85%	Singapore	2.97%
9	Macau	2.19%	India	2.34%	India	2.49%	India	2.69%	U.K.	2.01%
10	India	2.15%	Vietnam	1.76%	Germany	1.80%	U.K.	2.01%	India	1.71%
11	Vietnam	1.81%	Germany	1.68%	U.K.	1.74%	Germany	1.98%	Germany	1.68%
12	Brazil	1.69%	Macau	1.64%	Macau	1.68%	Bangladesh	1.61%	France	1.30%
13	Germany	1.68%	U.K.	1.61%	Vietnam	1.50%	Australia	1.60%	Vietnam	1.17%
14	Indonesia	1.66%	Indonesia	1.53%	France	1.42%	France	1.56%	Indonesia	1.13%
15	Canada	1.57%	Australia	1.39%	Australia	1.31%	Macau	1.47%	Australia	1.10%

*Compiled by TCA based on data publicized by the Ministry of Internal Affairs and Communications

2-2-5-2-2 Trends in Share of Incoming Call Duration by Country

Ranking	FY2016		FY2017		FY2018		FY2019		FY2020	
1	Korea	16.66%	U.S.A. (mainland)	18.75%	China	22.43%	China	25.12%	U.S.A. (mainland)	27.52%
2	U.S.A. (mainland)	15.06%	China	18.50%	U.S.A. (mainland)	20.30%	U.S.A. (mainland)	20.12%	Korea	27.40%
3	China	13.28%	Korea	12.60%	Korea	18.48%	Korea	18.92%	China	26.51%
4	Hong Kong	6.03%	Hong Kong	8.82%	Hong Kong	12.73%	Hong Kong	14.03%	Hong Kong	3.17%
5	Taiwan	5.09%	Taiwan	4.26%	Canada	2.33%	Canada	3.16%	Canada	2.05%
6	Luxembourg	4.57%	Luxembourg	3.29%	Singapore	2.14%	Singapore	2.45%	Australia	1.62%
7	Singapore	3.71%	Germany	2.87%	Luxembourg	1.75%	Taiwan	1.23%	Singapore	1.57%
8	Indonesia	3.20%	Thailand	2.83%	France	1.73%	Australia	1.20%	Germany	1.38%
9	Thailand	3.20%	France	2.70%	Taiwan	1.70%	Germany	1.15%	Thailand	0.96%
10	U.K.	3.15%	Singapore	2.69%	Germany	1.66%	Macau	1.08%	Taiwan	0.91%
11	Macau	2.91%	Canada	2.68%	Malaysia	1.48%	Malaysia	1.06%	Malaysia	0.86%
12	Germany	2.81%	U.K.	2.12%	Thailand	1.47%	Thailand	1.05%	Belgium	0.77%
13	Malaysia	2.03%	Belgium	1.95%	Macau	1.30%	France	0.89%	U.K.	0.66%
14	Belgium	2.01%	Indonesia	1.90%	Indonesia	1.24%	Iceland	0.77%	UAE	0.54%
15	France	1.91%	Malaysia	1.58%	Australia	1.11%	Indonesia	0.74%	Vietnam	0.53%

*Compiled by TCA based on data publicized by the Ministry of Internal Affairs and Communications

2-2-5-2-3 Outgoing and Incoming Call Duration by Country (FY2020)

Country (descending order according to outgoing duration)	Outgoing from Japan						Incoming to Japan					
	Ranking in outgoing		Volume of outgoing (Million minutes)	Increase or decrease ratio over previous year (%)	Share (%)	Accumulated share (%)	Ranking in incoming		Volume of outgoing (Million minutes)	Increase or decrease ratio over previous year (%)	Share (%)	Accumulated share (%)
	2020	2019					2020	2019				
U.S.A. (mainland)	1	(1)	90.8	▲7.76%	35.13%	35.13%	1	(2)	145.1	9.06%	27.52%	27.52%
China	2	(3)	41.8	▲48.90%	16.15%	51.28%	3	(1)	139.8	▲15.83%	26.51%	54.04%
Hong Kong	3	(2)	22.9	▲75.97%	8.86%	60.14%	4	(4)	16.7	▲81.99%	3.17%	57.20%
Korea	4	(4)	16.2	▲36.83%	6.26%	66.40%	2	(3)	144.4	15.47%	27.40%	84.60%
Thailand	5	(5)	9.1	▲47.60%	3.51%	69.92%	9	(12)	5.0	▲27.16%	0.96%	85.56%
Philippines	6	(6)	9.0	▲45.69%	3.49%	73.40%	18	(23)	1.8	▲23.57%	0.34%	85.91%
Taiwan	7	(7)	8.3	▲44.75%	3.20%	76.60%	10	(7)	4.8	▲40.99%	0.91%	86.82%
Singapore	8	(8)	7.7	▲45.69%	2.97%	79.57%	7	(6)	8.3	▲48.90%	1.57%	88.39%
U.K	9	(10)	5.2	▲47.94%	2.01%	81.58%	13	(16)	3.5	▲22.46%	0.66%	89.05%
India	10	(9)	4.4	▲66.79%	1.71%	83.29%	20	(20)	1.0	▲72.44%	0.18%	89.23%
Germany	11	(11)	4.4	▲55.71%	1.68%	84.98%	8	(9)	7.3	▲3.81%	1.38%	90.61%
France	12	(14)	3.4	▲56.73%	1.30%	86.27%	17	(13)	1.9	▲68.31%	0.35%	90.97%
Vietnam	13	(16)	3.0	▲56.30%	1.17%	87.44%	15	(17)	2.8	▲36.60%	0.53%	91.50%
Indonesia	14	(17)	2.9	▲52.97%	1.13%	88.57%	16	(15)	2.7	▲45.78%	0.50%	92.00%
Australia	15	(13)	2.8	▲64.44%	1.10%	89.67%	6	(8)	8.5	7.30%	1.62%	93.62%
Malaysia	16	(18)	2.5	▲46.06%	0.95%	90.62%	11	(11)	4.5	▲35.69%	0.86%	94.48%
Canada	17	(20)	2.4	▲35.60%	0.91%	91.53%	5	(5)	10.8	▲48.18%	2.05%	96.54%
Hawaii (U.S.A.)	18	(19)	2.4	▲41.92%	0.91%	92.45%	19	(24)	1.5	▲7.30%	0.28%	96.82%
Belgium	19	(29)	1.3	▲4.55%	0.52%	92.96%	12	(19)	4.1	14.49%	0.77%	97.59%
Pakistan	20	(44)	1.1	134.64%	0.41%	93.37%	33	(41)	0.2	▲40.96%	0.04%	97.63%
Brazil	21	(22)	1.0	▲59.12%	0.40%	93.77%	26	(30)	0.6	▲25.39%	0.11%	97.74%
Italy	22	(21)	0.9	▲69.54%	0.35%	94.12%	27	(29)	0.6	▲29.26%	0.11%	97.85%
Netherlands	23	(25)	0.8	▲57.08%	0.29%	94.41%	31	(38)	0.3	▲36.39%	0.05%	97.90%
New Zealand	24	(26)	0.7	▲57.71%	0.28%	94.69%	25	(22)	0.7	▲72.93%	0.13%	98.02%
Nepal	25	(24)	0.7	▲61.14%	0.27%	94.96%	62	(80)	0.0	▲48.72%	0.01%	98.03%
UAE	26	(27)	0.7	▲57.10%	0.27%	95.23%	14	(18)	2.9	▲23.82%	0.54%	98.57%
Sri Lanka	27	(28)	0.6	▲57.21%	0.25%	95.48%	22	(26)	0.7	▲37.43%	0.14%	98.71%
Mexico	28	(33)	0.6	▲46.56%	0.24%	95.71%	22	(25)	0.8	▲36.09%	0.14%	98.86%
Spain	29	(31)	0.6	▲58.29%	0.22%	95.93%	32	(34)	0.2	▲46.92%	0.05%	98.90%
Myanmar	30	(34)	0.6	▲41.49%	0.22%	96.16%	29	(36)	0.3	▲25.53%	0.07%	98.97%
Total of other countries	—	—	9.9		3.84%	100.00%	—	—	5.4		1.03%	100.00%
Total of all countries	—	—	258.5		—	—	—	—	527.1		—	—

*Compiled by TCA based on data publicized by the Ministry of Internal Affairs and Communications

2-3 Movements of Services and Charges

2-3-1 Fixed Telephones

2-3-1-1 Progress of Rates

2-3-1-1-1 Progress of Telephone Rates of NTT

1985	A three-minute call to the longest distance zone covering over 320km cost ¥400.
July 1986	First reduction of rates after NTT privatization was implemented. The Saturday discount was introduced, which applied, as was the case with holidays and nighttime, 40% discount from the normal rates for weekdays.
February 1988	NTT reduced the longest distance rate for weekday daytime calls to a level of ¥360 for 3 minutes.
February 1989	NTT reduced the longest distance rate for weekday daytime calls to a level of ¥330 for 3 minutes. It also cut rates for calls to the adjacent distance zone and areas within a radius of 20km from a level of ¥30 to ¥20 for 3 minutes (First reduction for short-distance calls since 1972).
March 1990	NTT reduced the longest distance rate for weekday daytime calls to a level of ¥280 for 3 minutes. It also introduced late-night discounts for local, short- and middle-distance calls.
March 1991	Distance segments covering over 160km were consolidated into a single longest distance zone, and the longest distance rate for weekday daytime calls was reduced to a level of ¥240 for 3 minutes. NTT also reduced rates for weekday daytime calls to areas within 20-30km radiuses to a level of ¥40 for 3 minutes. Late-night discount time period was extended by two hours to cover from 11 p.m. to 8 a.m. in the next morning.
June 1992	NTT reduced the longest distance rate for weekday daytime calls to a level of ¥200 for 3 minutes.
October 1993	NTT streamlined the distance segments covering 30-100km to two from four steps, and reduced rates for portions exceeding 30km by ¥10-60. The longest distance rate for weekday daytime calls was reduced to a level of ¥180 for 3 minutes.
March 1996	The rate for longest distance calls was lowered to ¥140 per 3 minutes in the daytime on weekdays.
February 1997	The rate for long-distance calls over 100 km was lowered to ¥110 per 3 minutes in the daytime on weekdays.
February 1998	Distances of over 100km were incorporated into the longest distance rate zone, and the longest distance rate for weekday daytime calls was reduced to a level of ¥90 for 3 minutes.
July 1999	With the reorganization of NTT, NTT East and NTT West took charge of intra-prefecture calls, and NTT Communications took inter-prefectures calls.
October 2000	NTT East and West lowered the toll call rate over 20 km in distance. The rate per 3 minutes in the daytime on weekdays was lowered to ¥30 for 20 to 60 km, and to ¥40 for over 60 km.
January 2001	NTT East reduced the local call rates to ¥9 per 3 minutes.
May 2001	NTT East and West lowered the local call rate to ¥8.5 per 3 minutes both in the daytime and at night.

2-3-1-1-2 Progress of Rates of Long-Distance and International NCCs

September 1987   Three new long-distance carriers stated services.
DDI CORPORATION, JAPAN TELECOM CO., LTD. and Teleway Japan Corporation started services. They offer charges about 25% below those of NTT. A 3-minute weekday daytime call to the longest distance zone of 340km cost ¥300 (in the case of NTT-established local portion charge being ¥20).

February 1988   These NCCs reduced evening and late-night rates, and introduced evening discounts into short-distance rates.

February 1989   Rates applicable to all the distance zones were reduced, bringing the longest distance rate for weekday daytime calls down to a level of ¥280 for 3 minutes.

March 1990   The longest distance rate for weekday daytime calls was reduced to a level of ¥240 for 3 minutes. Rates for calls to all the distance zones for evening, Saturdays, Sundays and holidays were reduced.

March 1991   Distance zones covering over 170km were consolidated into the longest distance zone, and the longest distance rate for weekday daytime calls was reduced to a level of ¥200 for 3 minutes. Evening, Saturday, Sunday, and holiday rates were also reduced.

April 1992   The longest distance rate for weekday daytime calls was reduced to a level of ¥180 for 3 minutes.

November 1993   In response to the introduction of the end-to-end charging (that was established by NCCs on an end-to-end basis for the entirety from the calling party through the called party including the local portion) in place of the add-on charging so far applied (total of the charge for trunk portion established by NCCs, and the charge for local portion established by NTT), an overall reduction of rates was implemented. As a result, the longest distance rate for weekday daytime calls was reduced to a level of ¥170 for 3 minutes.
The late-night discount time zone (from 11 p.m. to 8 a.m. in the next morning) was established, and the distance zones covering from 60km up to 100km were combined from two to one.

April 1994   The charge for the end portion provided by NTT was changed from the user charge to the cost-based inter-carriers settlement charge (access charge).

March 1996   In response to the reduction of the inter-carrier settlement charges paid by NCCs to NTT relating to the local portion provided by NTT (so-called "access charge"), the longest distance (over 170km) rate for weekday daytime calls was reduced to a level of ¥130 for 3 minutes from ¥170. In addition, the distance zone for short-distance calls, which had been set up in terms of "up to 60km" was divided into two zones, "up to 30km" and "over 30km up to 60km", and the rate-cut was made for "up to 30km" weekday daytime calls, and "up to 30km" and "over 30km up to 60km" late-night and early morning calls.

February 1997   The longest distance rate for weekday daytime calls was reduced to a level of ¥100 for 3 minutes.

February 1998   The longest distance rate for weekday daytime calls was reduced to a level of ¥90 for 3 minutes (The reduction brought NCCs' rates to the same level as NTT's).

The distance zones for the adjacent zone and the inside radius of 20km were established.

July 1998   KDD made a full-scaled inroad into domestic telephone markets, setting the longest distance rate for weekday daytime calls at a level of ¥69 for 3 minutes.

April 2000   Daytime and evening rates, etc to 20 - 30km and 30 - 60km distance zones were reduced
NTT Communications reduced daytime and evening rates for calls to 30 - 60km and 60 - 100km distance zones, and evening and midnight rates for 60 - 100km and over 100km distance zones.

October 2000   KDD, DDI and IDO merged into KDDI. New Intra-prefecture rates were established at a level of ¥40 for 3-minute weekday daytime call to the 60km or longer distance zone.

December 2000	C&W IDC fully entered the local domestic telephone market, and started the service setting, at a level of ¥45, its remotest distance rate applicable to 3-minute calls of 100km or longer distances for all day.
March 2001	The rate to the remotest distance zone was reduced to a level of ¥80 for 3-minute weekday daytime call, and the rate applicable to the 60-100km distance zone to a level of ¥60 for 3-minute weekday daytime call. NTT Communications reduced rates applicable to the 20 - 30km distance zone for all day, the 30 - 60km distance zone during evening and midnight, the 60 - 100km distance zone during midnight, and the more than 100km distance zone during daytime and midnight.
April 2001	Fusion Communications started IP telephone service, establishing its rate at ¥20 for 3-minute irrespective of distance throughout Japan.
May 2001	NTT Communications entered the local call market in Tokyo, Aichi, and Osaka. The rate is ¥8.5 per 3 minutes. KDDI and Japan Telecom entered the local call market. Their local call rate is ¥8.5 for 3-minute weekday daytime call.
December 2004	Japan Telecom started "OTOKU Line" fixed telephone service.
February 2005	KDDI started "Metal Plus" telephone service.
June 2006	Japan Telecom Co. Ltd. took over telecommunications business from Heisei Denden Corp. and Heisei Denden Communications Corp.
October 2006	Japan Telecom Co. Ltd. changed its company name to SoftBank Telecom Corp.
April 2015	SoftBank Mobile Corp., SoftBank BB Corp., SoftBank Telecom Corp., and Ymobile Corporation merged together to form SoftBank Mobile Corp.
July 2015	SoftBank Mobile Corp. changed its company name to SoftBank Corp.
December 2015	Fusion Communications Corp. changed its company name to Rakuten Communications Corp.
June 2016	KDDI terminated its "Metal Plus" telephone service.
July 2019	Rakuten Communications Corp. transferred its domestic telephone service (MYLINE) and the Rakuten Denwa phone service to Rakuten Mobile, Inc. through a company split.

(Reference) Carriers Participating in MYLINE

(As of October 2022)

Carrier / Call Catogory	ID number of telephone company	Local	Intra-pref long-distance	Outside of Prefecture	International	Registration available in
NTT East CORPORATION	0036	○	○			Eastern Japan
NTT West CORPORATION	0039	○	○			Western Japan
NTT Communications Corporation	0033	○	○	○	○	Nationwide
KDDI CORPORATION	0077 001 (International call)	○	○	○	○	Nationwide
SoftBank Corp.	0088 0061 (International call)	○	○	○	○	Nationwide
Rakuten Mobile, Inc.	0038	○	○	○	○	Nationwide
ARTERIA Networks Corporation	0060	○	○	○	○	Tokyo and 17 prefectures

*MY LINE Website : http://www.myline.org/index_e.html

2-3-1-1-3 Progress of Rates of Regional and Cable TV Operators

May 1988	Tokyo Telecommunication Network Company Inc. (called TTNet hereafter, later reformed to the present Poweredcom), a regional common carrier, started direct subscriber telephone service.
June 1997	Cable TV operator, TITUS COMMUNICATIONS CORPORATION, started subscriber telephone services. For call billing the Hudson charging method in units of 20 seconds was introduced.

July 1997	Suginami Cable TV Co., Ltd. (currently J-COM Tokyo) started subscriber telephone services.

July 1997 — Suginami Cable TV Co., Ltd. (currently J-COM Tokyo) started subscriber telephone services.

January 1998 — TTNet started relay telephone services with the rate of ¥9 for 3 minutes intra-zone calls, and the longest distance rate set at ¥72 for 3 minutes on weekday daytime calls.

March 1998 — TTNet reduced the longest distance rate for weekday daytime calls to a level of ¥63 for 3 minutes.

April 1999 — Kyushu Telecommunication Network Co., LTD. (hereafter, QTNet) started relay telephone services with the rate of ¥9 for intra-zone calls for 3 minutes on weekdays during the daytime, and ¥70 for the longest distance.

May 2000 — TTNet reduced the rate for 3-minute weekday daytime call to 60 - 100km distance zone from ¥54 to ¥45.

November 2000 — QTNet established new intra-prefecture rate, setting weekday daytime rate for call to a 60km or longer distance zone at a level of ¥27 for 3-minute.

May 2001 — TTNet reduced the charges for calls to all the distance zones. The charge applicable to the remotest distance zone was reduced to a level of ¥54 for 3-minute daytime call, ¥36 for 3-minute daytime call to a 60 -100km distance zone, and ¥8.4 for local calls, respectively.
QTNet reduced the rate for intra-zone calls to ¥8.4 for three minutes during the day on weekdays.

April 2003 — POWEREDCOM merged with TTNet, and the new company was named POWEREDCOM, Inc.

July 2004 — The telephone business of POWEREDCOM is merged with FUSION COMMUNICATIONS CORP.

June 2018 — QTnet (formerly Kyushu Telecommunication Network) terminated its relay telephone services.

April 2019 — K-Opticom Corporation changed its company name to OPTAGE Inc.

2-3-1-1-4 Progress of ISDN Service Provision

April 1988 — NTT inaugurated ISDN service.

October 1995 — Osaka Media Port and Shikoku Information and Telecommunication Network inaugurated ISDN service.

February 1996 — NTT started "INS Telehodai", a fixed rate service to selected telephone numbers in the midnight to early morning time zone.

March 1996 — HOKKAIDO TELECOMMUNICATION NETWORK and Tohoku Intelligent Telecommunication inaugurated ISDN service.

April 1996 — Chubu Telecommunications inaugurated ISDN service.

April 1997 — TTNet and QTNet inaugurated ISDN service.

July 1997 — NTT inaugurated ISDN service free of the facilities installation charge, "INS Net 64 Lite".

October 1997 — Chugoku Telecommunication Network inaugurated ISDN service.

December 1997 — Osaka Media Port started interconnection with NTT.

July 2000 — NTT East and NTT West inaugurated fixed rate IP connection service, "FLET'S ISDN".

July 2003 — Chugoku Telecommunication Network merged with Chugoku Information System Service and reorganized as Energia Communications.

April 2010 — Tohoku Intelligent Telecommunication terminated ISDN service.

March 2011 — Energia Communications terminated ISDN service.

December 2013 — QTNet terminated its ISDN service.

• Changes in NTT's Call Rates (for a 3-minute weekday daytime call)

Time of Revision	Number of Distance Zone	Within Zone	Adjacent Zone up to 20km	-30km	-40km	-60km	-80km	-100km	-120km	-160km	-240km	-320km	-500km	-750km	Over 750km
Before Aug. 1983	14	10	30	50	60	90	120	140	180	230	280	360	450	600	720
Aug. 1983	14	10	30	50	60	90	120	140	180	230	280	360	450	520	600
Jul. 1985	12	10	30	50	60	90	120	140	180	230	280	360	400		
Jul. 1986	10	10	30	50	60	90	120	140	180		260		400		
Feb. 1988	10	10	30	50	60	90	120	140	180		260		360		
Feb. 1989	10	10	30	50	60	90	120	140	180		260		330		
Mar. 1990	10	10	30	50	60	90	120	140	180		260		280		
Mar. 1991	9	10	30	40	60	90	120	140	180		240				
Jun. 1992	9	10	30	40	60	90	120	140	180		200				
Oct. 1993	7	10	30	40	50		80		140		180				
Mar. 1996	6	10	30	40	50		80		140						
Feb. 1997	6	10	30	40	50		80		110						
Feb. 1998	6	10	30	40	50		80		90						
(inter-Pref.) NTT Com — Apr. 2000	–	–	20	40			70				90				
(inter-Pref.) NTT Com — Mar. 2001	–	–	20	40			60				80				
(intra-Pref.) NTT East & West — Oct. 2000	–	10	20	30			40								
(intra-Pref.) NTT East & West — Jan. 2001	–	9 *	20	30			40								
May. 2001	–	8.5	20	30			40								

Shadowed columns are revised.   *In January 2001 only NTT East reduced the local call rates.

[Discout System by Day of the Week and Time Zone]

Nov. 1980	• Expansion of evening discount system • Establishment of midnight discount system [• 60% discount for calls to more-than-320km zones] [• 9p.m.- 6a.m.]
Aug. 1981	• Establishment of Sunday/Holiday discount system [• 40% discount for Sunday/Holiday daytime calls to more-than-60km zones]
Jul. 1986	• Establishment of Saturday discount system [• 40% discount for Saturday daytime calls to more-than-60km zones]
Mar. 1990	• Expansion of midnight discount system [• 25% discount for intra-zone and short-distance calls] [• 45% discount for medium- and long-distance calls] [• 11p.m. - 6a.m.]
Mar. 1991	• Expansion of midnight discount system [• 11p.m. - 8a.m.]
Oct.1993	• Expansion of midnight discount rate [• 50 - 55% discount for medium- and long-distance calls]
Oct. 2000	• Expansion of midnight discount system [• 20% discount for calls to 20 - 60km section]

2-3-2 Mobile Phones and PHS Services

2-3-2-1 Progress of Service Provision and Movements of Carriers — Mobile Phones

December 1979	NTT Public Corp. inaugurated automobile telephone service in 23 Tokyo Metropolitan wards.
April 1987	NTT inaugurated cellular telephone service.
December 1988	Nippon Idou Tsushin Corp. (IDO) inaugurated mobile services based on the NTT large-capacity system.
July 1989	KANSAI CELLULAR TELEPHONE COMPANY inaugurated mobile services based on the TACS system.
July 1992	NTT split up its mobile communications business division, establishing NTT Mobile Communications Network, Inc. (NTT DOCOMO).
March 1993	NTT DOCOMO inaugurated mobile services based on the 800MHz band PDC system.
July 1993	NTT DOCOMO was regionally divided into 9 regional companies under the one-region-one-company system.
October 1993	NTT DOCOMO abolished the deposit money (¥100,000) system.
April 1994	The mobile terminal COAM (Customer Owned and Maintained) system was introduced. Tokyo Digital Phone Co., Ltd. and TU-KA Phone Kansai Co., Ltd. inaugurated mobile services based on the 1.5GHz band PDC system. NTT DOCOMO inaugurated mobile services based on the 1.5GHz band PDC system in Tokyo Metropolitan 23 wards.
June 1994	IDO inaugurated mobile services based on the TACS system.
January 1996	Digital TU-KA Kyushu Co., Ltd. inaugurated mobile services based on the 1.5GHz band PDC system.
December 1996	The prior notification system of mobile communications rate was started. The new subscription fee was abolished.
March 1997	NTT DOCOMO inaugurated packet communications service, "DoPa."
July 1998	DDI Cellular Group started "cdmaOne" service in Kansai, Kyushu and Okinawa.
October 1998	TU-KA Phone Kansai Co., Ltd. inaugurated prepaid cellular telephone service.
January 1999	The 11-digit numbering system was introduced to the mobile telephone service.
February 1999	NTT DOCOMO inaugurated Internet connection service, "i-mode."
March 1999	NTT DOCOMO and IDO terminated mobile services based on the NTT large-capacity system.
April 1999	DDI Cellular Group and IDO extended service areas of "cdmaOne" to cover the whole nation, and inaugurated Internet connection service, "EZweb/EZaccess."
December 1999	J-Phone Group inaugurated Internet connection service, "J-Sky."
January 2000	DDI Cellular Group and IDO inaugurated packet communications service, "PacketOne."
April 2000	DDI Cellular Group and IDO started international roaming service "GLOBAL PASSPORT".
September 2000	DDI Cellular Group and IDO terminated mobile services based on the TACS system.
October 2000	DDI, KDD and IDO merged as DDI CORPORATION (KDDI). Nine J-Phone Group companies are merged for reorganization into J-Phone East Co., Ltd., J-Phone Central Co., Ltd. and J-Phone West Co., Ltd.
November 2000	Seven companies excluding OKINAWA CELLULAR TELEPHONE of DDI Cellular Group merged as au Corp.
October 2001	KDDI merged with au.
October 2001	NTT DOCOMO started full-scale services for IMT-2000 based on the W-CDMA system.

November 2001	J-Phone Co. Ltd. as the holding company merged with J-Phone East Co., Ltd., J-Phone Central Co., Ltd. and J-Phone West Co., Ltd., and the new company was named J-Phone Co., Ltd.
November 2001	KDDI and Okinawa Cellular Telephone Company launched the cellular telephone with GPS navigation function for the first time in the Japanese market.
April 2002	KDDI and Okinawa Cellular Telephone Company started CDMA2000 1x service.
December 2002	J-Phone Co., Ltd. started 3G service using 3GPP-based W-CDMA system, and international roaming with GSM-based networks.
June 2003	NTT DOCOMO started international roaming with GSM-based networks.
October 2003	J-Phone Co., Ltd. was renamed as Vodaphone K.K.
October 2003	Vodafone inaugurated "Vodafone live!" as the 3G Internet connection service, which is also available at overseas locations.
November 2003	KDDI and Okinawa Cellular Telephone Company launched CDMA 1X WIN service.
January 2004	NTT DOCOMO inaugurated "i mode Disaster Message Board Service".
May 2004	KDDI and Okinawa Cellular Telephone launched CDMA-based international data roaming services.
July 2004	NTT DOCOMO started to provide "i-mode FeliCa" service.
October 2004	Vodafone Holdings K.K. and Vodafone K.K. were merged into new Vodafone K.K.
December 2004	Vodafone launched international video telephone roaming services.
December 2004	NTT DOCOMO launched W-CDMA type 3G mobile network services based on 3GPP, packet roaming services with GSM (GPRS) networks to make overseas i-mode connection possible, and international video telephone roaming services.
September 2005	KDDI and Okinawa Cellular Telephone Company started to provide "EZ FeliCa" service.
September 2005	KDDI and Okinawa Cellular Telephone Company started au IC card service and international roaming with GSM-based networks.
September 2005	Vodaphone started 3G data card international roaming service.
September 2005	NTT DOCOMO started to provide the "i-channel" service based on "Flash Cast".
October 2005	KDDI merged with three Tu-Ka companies.
October 2005	Vodaphone launched "Vodaphone live! NAVI", a new navigation service allowing use of network-assisted GPS function not only in Japan but also abroad.
November 2005	Vodaphone started to provide "Vodaphone live! NAVI".
November 2005	NTT DOCOMO started to provide "Push-talk" voice communication service making use of the packet network.
November 2005	KDDI and Okinawa Cellular Telephone Company started "Hello Messenger" service.
November 2005	EMOBILE Ltd. received a radio frequency license for the 1.7GHz frequency band from the Ministry of Internal Affairs and Communications and entered into mobile phone business based on the W-CDMA system.
December 2005	KDDI and Okinawa Cellular Telephone Company launched the terminal compatible with "One-Seg" ground digital telecasting service for mobile and cellular telephones.
December 2005	NTT DOCOMO started to provide a new mobile credit brand "iD".
January 2006	KDDI and Okinawa Cellular Telephone started to provide "au LISTEN MOBILE SERVICE (LISMO)".
March 2006	NTT DOCOMO launched mobile telephone terminals conforming to the one segment terrestrial digital TV service.
April 2006	NTT DOCOMO started to provide "DCMX" credit service.
April 2006	Vodafone joined the SoftBank group.
May 2006	Vodafone released a cellular phone terminal conforming to the one-segment terrestrial

	digital TV service.
August 2006	NTT DOCOMO launched "HSDPA" conforming to high-speed packet communications.
	NTT DOCOMO started to provide the "music channel" service.
September 2006	KDDI and Okinawa Cellular started "EZ Channel Plus" and "EZ News Flash" utilizing the "BCMCS".
October 2006	Vodafone changed its company name to SoftBank Mobile Corp. SoftBank Mobile started a new portal site "Yahoo! Keitai". SoftBank Mobile launched "3G high speed".
October 2006	Three cellular phone companies started a mobile number portability system.
December 2006	KDDI and Okinawa Cellular Telephone started "EV-DO Rev.A" service.
March 2007	EMOBILE started the "EM mobile broadband" HSDPA data communication service.
May 2007	NTT DOCOMO started to provide the "2in1" service, where a single mobile phone unit has the functions of two mobile phone units.
December 2007	NTT DOCOMO started to provide the "Area Mail" service.
March 2008	KDDI terminated its Tu-Ka service. KDDI and Okinawa Cellular Telephone Company started GSM-based international data-roaming service.
March 2008	EMOBILE started voice communication service based on W-CDMA, and the "EMnet" internet connection service for cellular phone terminals.
June 2008	NTT DOCOMO started to provide the "Home U" service, which allows the use of mobile phones in a broadband environment such as in the home.
July 2008	SoftBank Mobile started to provide the "Double Number" service, which allows a single mobile phone unit to manage two phone numbers and e-mail addresses.
November 2008	EMOBILE started a High-Speed Uplink Packet Access (HSUPA) data communication service.
March 2009	SoftBank Mobile started a high-speed mobile data communication service for PCs.
July 2009	EMOBILE started a High-Speed Packet Access Plus (HSPA+) data communication service.
June 2010	KDDI inaugurated ISP for smartphones "IS NET".
September 2010	NTT DOCOMO inaugurated ISP for smartphones "sp mode".
December 2010	NTT DOCOMO inaugurated LTE high-speed data communication service with maximum 75Mbps download traffic speed "Xi (crossy) service".
December 2010	EMOBILE inaugurated high-speed packet communication service with maximum 42Mbps download traffic speed "EMOBILE G4".
February 2011	SoftBank Mobile inaugurated high-speed packet communication service with maximum 42Mbps download traffic speed "ULTRA SPEED".
March 2011	NTT DOCOMO and KDDI started to provide "Disaster Message Board Service" for smartphones.
April 2011	NTT DOCOMO inaugurated SIM unlock.
May 2011	eAccess started selling EMOBILE terminals with SIM unlock.
July 2011	Inter-carrier settlement for Short Message Service (SMS) is inaugurated.
January 2012	SoftBank Mobile began providing Disaster Info.
January 2012	KDDI began providing disaster and evacuation information through its Early Warning Mail services.
January 2012	KDDI began providing mobile NFC services.
February 2012	SoftBank Mobile began providing its "SoftBank 4G" high-speed data communication service with a maximum downstream speed of 110 Mbps.
February 2012	NTT DOCOMO began delivering early warning Area Mails (tsunami warnings).
March 2012	NTT DOCOMO began providing Disaster Voice Messaging Service.

March 2012	eAccess began providing its "EMOBILE LTE" high-speed data communication service with a maximum downstream speed of 75 Mbps.
March 2012	NTT DOCOMO began selling its "Mobacas" V-High multimedia broadcasting compatible terminals (first such attempt in Japan).
March 2012	KDDI began providing tsunami warnings in its Early Warning Mail services.
March 2012	NTT DOCOMO terminated its PDC service.
April 2012	KDDI introduced the EV-DO Advanced, a technology to ease data communication congestion at wireless base stations.
June 2012	KDDI began providing a Disaster Voice Messaging Service.
July 2012	SoftBank Mobile began providing a Disaster Voice Messaging Service.
July 2012	SoftBank Mobile began providing services using the 900 MHz band.
August 2012	SoftBank Mobile began providing tsunami warnings.
August 2012	Telecommunications carriers began "all-carrier search services" for mobile phone and PHS disaster message board services and NTT EAST/WEST Disaster Message Board (web171).
September 2012	KDDI began providing the 4G LTE service based on the next-generation high-speed communication standard, LTE (Long Term Evolution).
October 2012	Business alliance between SoftBank Mobile and eAccess.
February 2013	NTT DOCOMO, China Mobile and KT developed common requirements for NFC international roaming.
February 2013	SoftBank Mobile began providing its SoftBank satellite phone service.
March 2013	eAccess began providing emergency earthquake warnings, tsunami warnings, and disaster and evacuation information through its Early Warning Mail services.
March 2013	eAccess began providing the FeliCa service.
March 2013	NTT DOCOMO, KDDI, SoftBank Mobile, and eAccess began providing mobile phone services throughout the entire Toei Subway Lines.
April 2013	NTT DOCOMO, KDDI, Okinawa Cellular, and SoftBank Mobile enabled interoperability of the Disaster Voice Messaging Service across the four mobile phone carriers.
July 2013	NTT DOCOMO, KDDI, and SoftBank Mobile began providing the LTE service at Mt. Fuji.
September 2013	SoftBank Mobile began providing international LTE roaming services.
September 2013	KDDI began providing international LTE roaming services.
October 2013	KDDI adopted the IEEE802.11ac next-generation wireless LAN standard for its au Wi-Fi SPOT public wireless LAN services.
November 2013	NTT DOCOMO, KDDI, Okinawa Cellular, SoftBank Mobile, and eAccess began using mobile phone numbers starting with 070.
November 2013	NTT DOCOMO developed a multi-band indoor base station and antenna.
January 2014	Six mobile phone and PHS carriers enabled interoperability of the Disaster Voice Messaging Service across these carriers.
March 2014	NTT DOCOMO began providing international LTE roaming services.
April 2014	NTT DOCOMO, KDDI, Okinawa Cellular, and SoftBank Mobile began delivering information on the protection of the people using the early warning Area Mails and Early Warning Mail services.
May 2014	Six mobile phone and PHS carriers standardized the number and varieties of pictographs used in text messages, including SMS, exchanged between carriers.
May 2014	KDDI introduced Carrier Aggregation, an LTE-Advanced technology based on the next-generation high-speed communication standard LTE, with a maximum receiving speed of 150 Mbps for the first time in Japan.
May 2014	NTT DOCOMO released guidelines for video distribution utilizing the next-generation

	video compression technology, HEVC.
June 2014	eAccess Ltd. and Willcom, Inc. merged.
June 2014	NTT DOCOMO developed the world's first new SIM-based authentication mini device, called Portable SIM.
June 2014	NTT DOCOMO began providing Japan's first VoLTE call service.
July 2014	eAccess Ltd. changed its company name to Ymobile Corporation.
August 2014	Ymobile started its new Y!mobile service.
October 2014	Number portability between mobile and PHS phones began.
November 2014	NTT DOCOMO started Japan's first international outbound roaming service on a TD-LTE network.
December 2014	KDDI began providing the au VoLTE next-generation voice calling service, utilizing the 4G LTE network.
December 2014	SoftBank Mobile began providing voice communication services using the VoLTE technology, a technology that enables voice communication over the LTE high-speed data communication network.
March 2015	NTT DOCOMO began providing LTE-Advanced services under the name "PREMIUM 4G" with a maximum downlink of 225 Mbps, which was the fastest in Japan.
April 2015	SoftBank Mobile Corp., SoftBank BB Corp., SoftBank Telecom Corp., and Ymobile Corporation merged together.
May 2015	The revised SIM unlocking guidelines came into effect, and NTT DOCOMO, KDDI, and SoftBank Mobile began providing SIM unlocking services based on the new guidelines.
July 2015	SoftBank Mobile Corp. changed its company name to SoftBank Corp.
October 2015	NTT DOCOMO became Japan's first telecommunications carrier to provide an international VoLTE roaming service.
March 2016	NTT DOCOMO began providing services using the world's first network function virtualization (NFV) technology that can run Evolved Packet Core (EPC) software from multiple vendors on its commercial network.
June 2016	KDDI began providing international VoLTE roaming services.
September 2016	SoftBank began providing the world's first commercial service with Massive MIMO (spatial multiplexing technology).
March 2017	NTT DOCOMO began providing communication service with a maximum downlink of 682 Mbps by introducing two new technologies: 256 QAM and 4×4 MIMO.
September 2017	KDDI began providing communication service with a maximum downlink of 708 Mbps by introducing 265 QAM and 4×4 MIMO.
May 2018	NTT DOCOMO, KDDI, and SoftBank began providing the +Message service, a new service as an extension of SMS, based on the GSMA specifications.
June 2018	NTT DOCOMO, in collaboration with China Mobile, commercialized the world's first IoT multi-vendor eSIM solution based on the GSMA 3.1 specifications.
October 2018	NTT DOCOMO, SoftBank, and KDDI each began providing services for VoLTE interconnection between different carriers.
October 2019	Rakuten Mobile began providing commercial service with the world's first end-to-end fully virtualized cloud-native network.
March 2020	NTT DoCoMo, KDDI, and SoftBank each began providing communication service using the fifth-generation mobile communication system (5G).
April 2020	Rakuten Mobile launched full-scale mobile carrier service.
September 2020	Rakuten Mobile began providing communication service using the fifth-generation mobile communication system.
October 2020	KDDI completed its succession of UQ mobile's business.
March 2021	SoftBank launched a new online-only plan under the brand name "LINEMO."

March 2021   KDDI launched a new online-only plan under the brand name "povo."

March 2021   NTT DOCOMO launched a new online-only plan under the brand name "ahamo."

March 2022   KDDI and Okinawa Cellular Telephone Company terminated their CDMA 1X WIN and other services for au 3G mobile phones.

Note: The transmission speeds referred to in the chronology are those at the time of the introduction of the corresponding services by the relevant companies.

2-3-2-2 Progress of Service Provision and Movements of Carriers — PHS

July 1995	DDI TOKYO POCKET TELEPHONE, Inc. DDI HOKKAIDO POCKET TELEPHONE Inc., NTT Central Personal Communications Network Inc., and NTT Hokkaido Personal Communications Network Inc. inaugurated services. After October 1995, 7 companies of DDI POCKET TELEPHONE Group, 7 companies of NTT Personal Communications Network Group and 10 companies of ASTEL Group inaugurated services.
February 1997	The new subscription fee was abolished.
December 1998	Nine companies of NTT Personal Communications Network Group assigned their business to nine companies of NTT DOCOMO Group.
April 1999	ASTEL Tokyo Corporation was merged into Tokyo Telecommunication Network Co., Inc. NTT DOCOMO Group inaugurated 64kbps data communications service.
November 1999	ASTEL Hokkaido Corporation assigned its business to HOKKAIDO TELECOMMUNICATION NETWORK CO., Inc.
January 2000	Nine companies of DDI POCKET TELEPHONE Group were amalgamated as DDI POCKET Inc.
September 2000	ASTEL Tohoku Corporation assigned its business to Tohoku Intelligent Telecommunication Co., Inc.
November 2000	ASTEL Chubu and CHUBU TELECOMMUNICATIONS CO., INC. merged. ASTEL KANSAI CORPORATION assigned its business to K-Opticom Corporation.
April 2001	ASTEL KYUSHU assigned its business to Kyushu Telecommunication Network Co., Inc.
August 2001	DDI Pocket Inc. inaugurated fixed-rate data communication service.
October 2001	Astel Chugoku Corporation assigned its business to Chugoku Information System Service Co., Inc.
December 2001	Astel Hokuriku Corporation assigned its business to Hokuriku Telecommunication Network Co., Inc.
March 2002	Astel Shikoku Corporation assigned its business to Shikoku Information and Telecommunication Network Company, Incorporated.
April 2002	Shikoku Information and Telecommunication Network Company, Incorporated changed the company name to STNet Incorporated.
August 2002	Tokyo Telecommunication Network Company, Incorporated assigned its PHS business to Magic Mail Inc.
October 2002	Magic Mail Inc. was merged with Yozan Inc.
April 2003	NTT DOCOMO group inaugurated fixed-rate data communication service.
July 2003	Chugoku Telecommunication Network merged with Chugoku Information System Service and reorganized as Energia Communications.
November 2003	Kyushu Telecommunications Network Co., Inc. terminated their PHS telephone service.
March 2004	Hokkaido Telecommunications Network Co., Inc. terminated their PHS telephone service.
May 2004	Hokuriku Telecommunications Network Co., Inc. terminated their PHS telephone service.
September 2004	K-Opticom Corporation terminated the PHS voice telephone service out of their PHS services.
October 2004	DDI Pocket, Inc. became independent from the KDDI group.
December 2004	Energia Communications ceased to provide PHS voice telephone service out of their PHS services.
January 2005	Astel Okinawa transferred goodwill to WILLCOM Okinawa.

February 2005	DDI Pocket, Inc. changed the name to WILLCOM, Inc.
May 2005	STNet ceased to provide their PHS telephone service.
May 2005	Chubu Telecommunications Co., Inc. ceased to provide their PHS communication service.
May 2005	Willcom started "Willcom Teigaku Plan" fixed-rate service.
June 2006	YOZAN terminated its PHS telephone service.
December 2006	Tohoku Intelligent Telecommunication terminated its PHS telephone service.
October 2007	Energia Communications terminated PHS services.
January 2008	NTT DOCOMO Group terminated their PHS services.
December 2010	Willcom started the "Fixed Rate with Anyone" service.
September 2011	K-Opticom terminated its PHS service.
June 2014	Willcom merged with eAccess (eAccess Ltd.).
January 2021	SoftBank terminated its PHS service.

2-3-3 International Telephone Services

2-3-3-1 Progress of Service Provision and Movements of Carriers

• In October 1989, International Telecom Japan Inc. (ITJ) and International Digital Communications Inc. (IDC) introduced services with 23% lower rates than those of Kokusai Denshin Denwa Co.,Ltd. (KDD)

• From 1989 through 1996 KDD implemented rate reductions eight times, and ITJ and IDC five times, resulting in a steady shift toward less expensive rates.

October 1998	DDI Corporation (DDI) started international telephone services with the level of charge set at ¥240 for a daytime 3-minute call to U.S. MCI Worldcom Japan, Inc. (WCOM) started international telephone services with the level of charge set at ¥248 for a daytime 3-minute call to U.S.
December 1998	KDD reduced charges for calls to all destinations (230 countries and areas). The average reduction rate was about 10.6%. As the result, a daytime 3-minute call to U.S. cost ¥240. Japan Telecom (JT) reduced charges for calls to 28 destinations. The average reduction rate was about 8.6%. A daytime 3-minute call to U.S. cost ¥240. IDC reduced charges for calls to 23 destinations. The average reduction rate was about 9.0%. A daytime 3-minute call to U.S. cost ¥240. WCOM reduced charges. A daytime 3-minute call to U.S. cost ¥150.
January 1999	DDI reduced charges for calls to 25 destinations. The average reduction rate was about 8.4%. A daytime 3-minute call to U.S. cost ¥168. JT reduced charges for calls to 97 destinations. The average reduction rate was about 2.2%. IDC reduced charges for calls to 51 destinations. The average reduction rate was about 3.5%.
March 1999	DDI reduced charges for calls to 27 destinations, with a main target of reduction on calls placed during 23:00 to 08:00 of the following day. The average reduction rate was about 5.8%.
July 1999	Tokyo Telecommunication Network Co.,Inc. (TTNet) started international telephone services with the level of charge set at ¥168 for a daytime 3-minute call to U.S.
October 1999	JT reduced charges for all destinations (223 countries and areas). The average reduction rate was about 10.3%. A daytime 3-minute call to U.S. cost ¥180. Cable & Wireless IDC reduced charges for calls to 192 destinations. The average reduction rate was about 10.9%. A daytime 3-minute call to U.S. cost ¥180. NTT Communications Corp. started international telephone services with the level of charge set at ¥180 for a daytime 3-minute call to U.S.
November 1999	KDD reduced charges for calls to all destinations (231 countries and areas). The average reduction rate was about 11.1%. A daytime 3-minute call to U.S. cost ¥180. DDI reduced charges for calls to 38 destinations. The average reduction rate was about 8.4%. A daytime 3-minute call to U.S. cost ¥156. TTNet reduced charges for calls to 58 destinations. The average reduction rate was about 11%. A daytime 3-minute call to U.S. cost ¥132.
December 1999	KDD reduced charges for cellular/PHS-originated calls to all destinations (231 countries/areas). The average reduction rate was about 11.9%.
February 2000	KDD reduced charges for calls to 17 destinations (Taiwan, China, U.K., France, Germany, etc.). The average reduction rate was about 1.4%.
October 2000	DDI, KDD and IDO were merged as KDDI.
April 2001	Fusion Communications started international telephone services, establishing the all-time flat rate system. The charge for 3-minute calls to U.S. is ¥90.
September 2001	Fusion Communications Corporation reduced the charges for calls to all destinations (230 countries and areas). A three-minute call to the U.S. cost ¥45.

April 2003	POWEREDCOM merged with TTNet, and the new company was named POWEREDCOM, Inc.
July 2004	The telephone business of POWEREDCOM is merged with FUSION COMMUNICATIONS CORP.
October 2006	Japan Telecom Co. Ltd. changed its company name to SoftBank Telecom Corp.
April 2015	SoftBank Mobile Corp., SoftBank BB Corp., SoftBank Telecom Corp., and Ymobile Corporation merged together to form SoftBank Mobile Corp.
July 2015	SoftBank Mobile Corp. changed its company name to SoftBank Corp.
December 2015	Fusion Communications Corp. changed its company name to Rakuten Communications Corp.
July 2019	Rakuten Communications Corp. transferred its international telephone service to Rakuten Mobile, Inc. through a company split.

2-3-4 Leased Circuit and Data Transmission Services

2-3-4-1 Progress of Service Provision and Movements of Carriers

• Progress of Leased Circuit Service Provision

(NTT)

December 1997	NTT started "Digital Access 128" as short-distance economy service.
April 1998	NTT started "Digital Access 1500" service.
August 1998	NTT started "Digital Reach" as medium- and long-distance economy service.
December 1998	NTT started "ATM SHARE LINK" as partial band assurance type exclusively for ATM.
October 1999	NTT Communications started "Gigaway" service.
March 2000	NTT Communications started "Air Access" service.
April 2001	NTT East and West started "Digital Access 6000" service.
November 2001	NTT East started "Metro High Link" service.
June 2002	NTT East started "Super-high Link" service.
July 2002	NTT West started "Giga Data Link" service.
October 2002	NTT Communications started "EtherArcstream" service.
June 2004	NTT Communications started "GIGASTREAM" service.
December 2008	NTT Communications started "GIGASTREAM Premium Ether" service.
May 2011	NTT Communications started to provide "Arcstar Universal One".

(Long-Distance and International Carriers)

April 1998	KDDI (TWJ) started to provide leased circuit service for ATM.
October 1998	Long-distance and International NCCs started economy services.
September to October 1999	Long-distance and International NCCs acquired rate setting right and started end-to-end rate services.
January 2000	Global Access started domestic and international leased circuit service.
July 2000	Japan Telecom started domestic wide-band leased circuit service.
October 2002	Japan Telecom started international wide-band leased circuit service.

(Regional Carriers)

April 1997	Nine electric power companies started joint high-speed digital transmission service.
January 1998	TTNet started FDDI leased circuit service.
April 1998	TTNet started leased circuit service for ATM.
May 1998	Ten electric power companies completed nationwide linkage of high-speed digital transmission services.
October 1998	Nine electric power companies started linkage of ATM leased circuit services.
August 1999	Ten electric power companies completed nationwide linkage of economy services.
April 2001	TTNet started to provide "PeneLink (leased circuit)" (Ethernet leased circuit service).
September 2001	Keio Network Communications started to provide "Express-Ether" service.
April 2002	Osaka Media Port started Ether leased circuit service.
June 2002	Chubu Telecommunication started optical fiber leased circuit service.
April 2003	Osaka Media Port started Ether Network service (W-Link).

(Regional CATV)

April 2002   Katch Network started optical fiber leased circuit service.

December 2002   Himawari Network started optical fiber leased circuit service.

December 2002   My Television started regional LAN services.

• Progress of Data Transmission Service Provision

(NTT)

December 1996	NTT started OCN service.
August 1999	NTT Communications started to provide OBN (Open Business Network) service.
September 1999	NTT Communications started to provide "Arcstar Value Access" service.
May 2000	NTT East and West started to provide Wide LAN Service.
July 2000	NTT Communications started "Super VPN (current Arcstar IP-VPN)" service.
July 2000	NTT DOCOMO and NTT Communications jointly started to provide "RALS (Remote Access Line Service)".
September 2000	NTT East started to provide FLET's Office".
October 2000	NTT Communications started to provide "Broadband Access" service.
October 2000	NTT East and West started to provide "Mega Data Nets" service.
December 2000	NTT Communications started to provide "Giga Ether Platform" service.
January 2001	NTT Communications started to provide "Arcstar Global IP-VPN" service.
March 2001	NTT East started to provide "Metro Ether" service.
April 2001	NTT Communications started to provide "e-VLAN" service.
May 2001	NTT West started to provide "Urban Ether" service.
March 2002	NTT East started to provide "FLET's Group Access" service.
March 2002	NTT East started to provide "Super Wide LAN Service".
March 2002	NTT West started to provide "Wide LAN Plus" service.
March 2003	NTT East started to provide "FLET's Office Wide" service.
April 2003	NTT Communications started to provide "Super HUB" service.
May 2003	NTT Communications started to provide "FLEXGIGAWAY" service.
July 2003	NTT East started to provide "Flat Ether" service.
October 2003	NTT West started to provide "Flat Ether" service.
December 2003	NTT East started to provide the Smart Ether service.
June 2004	NTT Communications started to provide the "Group-VPN" service.
April 2006	NTT West started to provide the "Business Ether" service.
May 2006	NTT East started to provide the "Business Ether" service.
July 2009	NTT Communications started to provide the "Group-Ether" service.
May 2011	NTT Communications started to provide "Arcstar Universal One".

(Long-Distance and International Carriers)

April 1997	Long-distance and International NCCs sequentially started to provide computer network services.
April 1999	Japan Telecom started to provide international cell relay service.
April 2000	Japan Telecom started to provide Solteria (IP-VPN) service.
October 2000	KDDI started to provide ANDROMEGA IP-VPN service.
February 2001	Fusion Communications started to provide FUSION IP-VPN service.
October 2001	Japan Telecom started to provide "Wide-Ether" (wide-area LAN).
December 2001	Cable & Wireless IDC started to provide "High-speed Ethernet Service".

December 2001   KDDI started to provide "Ether-VPN" service.

September 2002   Cable & Wireless IDC started to provide "IP-VPN QoS" service.

November 2002   Japan Telecom started to provide "ASSOCIO (MLPS Traffic Switching Service)".

August 2012   SoftBank Telecom began providing its White Cloud SmartVPN service.

(Regional Carriers)

From September 1997   Power company based NCCs sequentially started to provide computer network services.

March 2001   Hokkaido Telecommunication Network, Inc started to provide wide-area Ethernet service "L2L".

April 2001   Poweredcom started to provide "Powered Ethernet" wide-area Ethernet connection service.

April 2001   TTNet started to provide "Pene-Link (Multi-access)" (wide-area Ethernet connection service).

June 2001   K-Opticom started to provide IP-VPN service.

July 2001   Poweredcom started to provide "Powered-IP MPLS" (IP-VPN connection service).

August 2001   Chugoku Telecommunication Network started to provide Ethernet communication network service "V-LAN".

June 2002   Keio Network Communications started to provide "Multi-Express Ether" service.

July 2003   Chugoku Telecommunication Network merged with Chugoku Information System Service and reorganized as Energia Communications.

January 2003   Chubu Telecommunication started to provide band-assured type Ether network service "CTC Ether Link".

June 2005   Chubu Telecommunication started to provide "CTC Ether DIVE" wide-area Ethernet service.

(Regional CATV)

December 1995   Himawari Network started to provide cell relay service.

November 1997   Katch Network started to provide cell relay service.

April 1998   MICS Network started to provide ATM switching service.

September 1999   MICS Network started to provide wide-area LAN service.

Chapter 3
Situation of TCA Members

Member List

*Symbols representing types of provided services

1. Subscriber telephone
2. ISDN (excluding switched telephone, public telephone and international digital communications service)
3. Switched telephone (excluding international telephone)
4-①. International telephone
4-②. International ISDN
5. Public telephone
6-①. Mobile phone (using a 3.9-generation mobile communication system)
6-②. Mobile phone (using a 5th generation mobile communication system)
6-③. Mobile phone (not using 3.9–4th generation mobile communication systems or a 5th generation mobile communication system)
7. PHS
8-①. IP phone (using 050/0AB-J number)
8-②. IP phone (not using 050/0AB-J number)
9. Wireless landline telephone
10. Satellite mobile communication service
11. FMC service
12. Internet connection service
13-①. FTTH access service
13-②. FTTH access service (using VDSL or other facility)
14. DSL access service
15. FWA access service
16. CATV access service
17. Mobile/PHS access service
18. 3.9-generation mobile phone access service
19. Fifth generation mobile communication access service
20. Local 5G service

21. Frame relay service
22. ATM switch service
23. Public wireless LAN access service
24-①. BWA service (nationwide BWA service)
24-②. BWA service (regional BWA service)
24-③. BWA service (private BWA service)
25. IP-VPN service
26. Wide-area Ethernet service
27. Satellite access service
28-①. Leased line service (domestic)
28-②. Leased line service (international)
29. Unlicensed LPWA service
30. Value added service using telecommunication services in 1 to 28 above
31. Internet-related service (excluding IP phone)
32-①. Mobile virtual network service (for mobile phones)
32-②. Mobile virtual network service (for PHS)
32-③. Mobile virtual network service (for local 5G service)
32-④. Mobile virtual network service (for BWA service)
33-①. Telecommunications service for domain name facilities (as specified in Article 59-2, paragraph (1), item (i)-(a))
33-②. Telecommunications service for domain name facilities (as specified in Article 59-2, paragraph (1), item (i)-(b))
33-③. Telecommunications service for domain name facilities (as specified in Article 59-2, paragraph (1), item (ii))
34-①. Telegram (involving acceptance and delivery jobs)
34-②. Telegram (without acceptance and delivery jobs)
35. Telecommunication services other than 1 through 33 above

* Representation of service areas

Hokkaido

Tohoku District: Aomori, Iwate, Miyagi, Akita, Yamagata, and Fukushima

Kanto District: Ibaraki, Tochigi, Gunma, Saitama, Tokyo, Chiba, Kanagawa, and Yamanashi

Shin-etsu District: Nagano and Niigata

Hokuriku District: Toyama, Ishikawa, and Fukui

Tokai District: Shizuoka, Aichi, Gifu, and Mie

Kinki District: Shiga, Kyoto, Osaka, Hyogo, Nara, and Wakayama

Chugoku District: Tottori, Shimane, Okayama, Hiroshima, and Yamaguchi

Shikoku District: Tokushima, Kagawa, Ehime, and Kochi

Kyushu District: Fukuoka, Saga, Nagasaki, Kumamoto, Oita, Miyazaki, and Kagoshima

Okinawa

*Compiled from the data in the questionnaire survey covering member companies based on the "Registered Member Companies" (Ministry of Internal Affairs and Communications).

As of July 1, 2022

Company	Service areas	1	2	3	4①	4②	5	6①	6②	6③	7	8①	8②	9	10	11	12	13①	13②	14	15	16	17	18	19	20	21	22	23	24①	24②	24③	25	26	27	28①	28②	29	30	31	32①	32②	32③	32④	33①	33②	33③	34①	34②	35
NIPPON TELEGRAPH AND TELEPHONE CORPORATION																																																		
NIPPON TELEGRAPH AND TELEPHONE EAST CORPORATION	Hokkaido, Tohoku District, Kanto District, and Shin-etsu District	O	O				O					O	O				O	O	O		O				O		O						O		O		O	O											O	O
NIPPON TELEGRAPH AND TELEPHONE WEST CORPORATION	Hokuriku District, Tokai District, Kinki District, Chugoku District, Shikoku District, Kyushu District, and Okinawa	O	O				O					O	O				O	O	O	O	O				O		O						O		O		O	O											O	O
KDDI CORPORATION	Nationwide		O	O	O		O	O		O		O	O		O		O	O	O					O	O	O			O				O	O	O	O	O		O	O	O			O			O		O	O
SoftBank Corp.	Nationwide	O	O	O	O	O		O	O	O	O		O		O		O	O	O				O	O	O	O			O	O			O	O	O	O	O		O	O	O			O						O
ARTERIA Networks Corporation	Nationwide		O									O	O				O	O	O														O	O		O	O	O	O	O					O					
NTT Communications Corporation	Nationwide		O	O	O							O	O				O	O	O														O	O		O	O												O	O
East Japan Railway Company	Nationwide																													O			O	O							O								O
SKY Perfect JSAT Corporation	Nationwide											O			O		O																O	O	O	O	O		O										O

| Company | Service areas | Types of services |||||||||||||||||||||||||||||||||||||||||||||||||
---	---	1	2	3	4①	4②	5	6①	6②	6③	7	8①	8②	9	10	11	12	13①	13②	14	15	16	17	18	19	20	21	22	23	24①	24②	24③	25	26	27	28①	28②	29	30	31	32①	32②	32③	32④	33①	33②	33③	34①	34②	35
PCCW Global (Japan) K.K.	Nationwide																○																					○													
Sony Network Communications Inc.	Nationwide											○	○				○	○	○												○	○		○			○	○	○	○			○								
Hokkaido Telecommunication Network Co., Inc	Hokkaido																○																	○	○				○												
	Tokyo																																						○												
Tohoku Intelligent Telecommunication Co.,Inc.	Tohoku District, and Niigata				○							○	○				○						○								○			○			○	○	○				○							
Hokuriku Telecommunication Network Co.,Inc	Hokuriku District											○					○														○	○			○	○			○	○										
Chubu Telecommunications Company, Incorporated	Tokai District, and Nagano											○	○		○		○	○	○												○	○		○	○		○	○	○				○							○
OPTAGE Inc.	Nationwide				○							○					○	○	○	○											○	○		○	○		○	○	○				○							○
Energia Communications, Inc.	Chugoku District											○	○				○	○	○															○			○	○	○											
STNet, Incorporated	Nationwide											○					○	○	○												○	○		○			○		○				○							○
QTnet, Inc.	Nationwide											○					○	○	○							○		○			○	○		○	○	○	○	○												
OTNet Company, Incorporated	Okinawa											○					○	○	○	○											○	○		○	○															
Japan Digital Serve Corporation	Kanto District																																	○	○				○											
Japan Network Engineering Co.,Ltd.	Nationwide											○	○		○	○	○																	○			○		○											
LCV Corporation	Shinetsu District											○					○	○	○			○						○	○					○					○											
Kintetsu Cable Network Co., Ltd.	Nara, Osaka, Kyoto, Mie, and Aichi											○					○	○	○			○						○	○					○																
	Nationwide																																													○			○	
its communications Inc.	Kanto District (Tokyo and Kanagawa)											○					○	○	○	○			○						○	○				○					○	○										
Cable Television Shinagawa inc.	Tokyo											○					○	○	○			○							○					○					○	○										
New Media. Corporation	Yonezawa City, Nan'yo City, Takahata Town, and Kawanishi Town in Yamagata Hakodate City, Hokuto City, and Nanae Town in Hokkaido Niigata City in Niigata Fukushima City in Fukushima											○					○	○	○			○						○	○					○					○										
CTY.co.,Ltd	Mie											○					○	○					○					○	○					○			○								○	○				

Types of services

Company	Service areas	1	2	3	4①	4②	5	6①	6②	6③	7	8①	8②	9	10	11	12	13①	13②	14	15	16	17	18	19	20	21	22	23	24①	24②	24③	25	26	27	28①	28②	29	30	31	32①	32②	32③	32④	33①	33②	33③	34①	34②	35
TOKYO CABLE NETWORK, INC.	Tokyo											O					O			O								O		O			O		O										O					
JCOM Co., Ltd.	Nationwide																O																		O			O	O	O					O					O
MICS NETWORK CORPORATION	Aichi	O																O	O			O						O		O	O	O																		
advanscope inc.	Tokai District							O										O	O			O						O	O																					
TOKAI Communications Corporation	Nationwide											O						O	O	O	O									O				O	O	O	O	O	O	O					O					O
Cable Networks AKITA	Tohoku District											O						O	O					O		O		O		O			O		O			O	O						O					
Matsusaka CATV Station Co.,Ltd. (MCTV)	Mie																	O	O		O							O					O					O												
COMMUNITY NETWORK CENTER INCORPORATED	Tokai District																											O		O				O				O	O											O
Igaueno Cable Television Co.,Ltd.	Mie										O							O	O	O		O						O		O				O	O															
ICHIHARA COMMUNITY NETWORK TELEVISION CORPORATION	Ichihara City, and Midori Ward, Chiba City										O							O	O	O		O	O							O					O				O	O	O									
CHUKAI CABLE TELEVISION SYSTEM OPERATOR	Western Tottori										O							O	O		O							O							O				O	O									
IRUMA CABLE TELEVISION CO., LTD	Kanto District										O						O			O								O							O				O	O	O									
NTT DOCOMO, INC.	Nationwide			O				O	O	O	O			O		O	O	O	O				O	O	O			O					O	O	O	O	O		O	O										
OKINAWA CELLULAR TELEPHONE COMPANY	Okinawa					O		O	O		O						O	O	O				O	O	O			O							O			O	O						O					
Rakuten Mobile, Inc.	Nationwide	O	O					O	O			O	O				O	O	O	O		O	O	O															O	O					O					O
Tokyo Telemessage Inc.	Nationwide																																																	O
AVICOM JAPAN CO., LTD.	Nationwide																																			O														
Kansai Airports Technical Services Co.,Ltd	Osaka																O																								O									
UQ Communications Inc.	Nationwide																O											O						O					O	O										O

Telecom Data Book 2022
(Compiled by TCA)
Planned / Edited / Published by Telecommunications Carriers Association

Koshin Bldg. 2F
1-10 Kanda-Ogawamachi, Chiyoda-ku, Tokyo 101-0052, Japan
Tel. 03-5577-5845 Fax. 03-5296-5520
https://www.tca.or.jp/
Co-edited / Printed HARIU Communications Co., Ltd.

Published: February 2023

Price: ¥2,200 (¥2,000 + 10% tax)
*Reproduction forbidden

Incorrectly collated books will be replaced
©2023 printed in Japan   ISBN978-4-906932-21-4